Places of interest

See Information pages
for further details

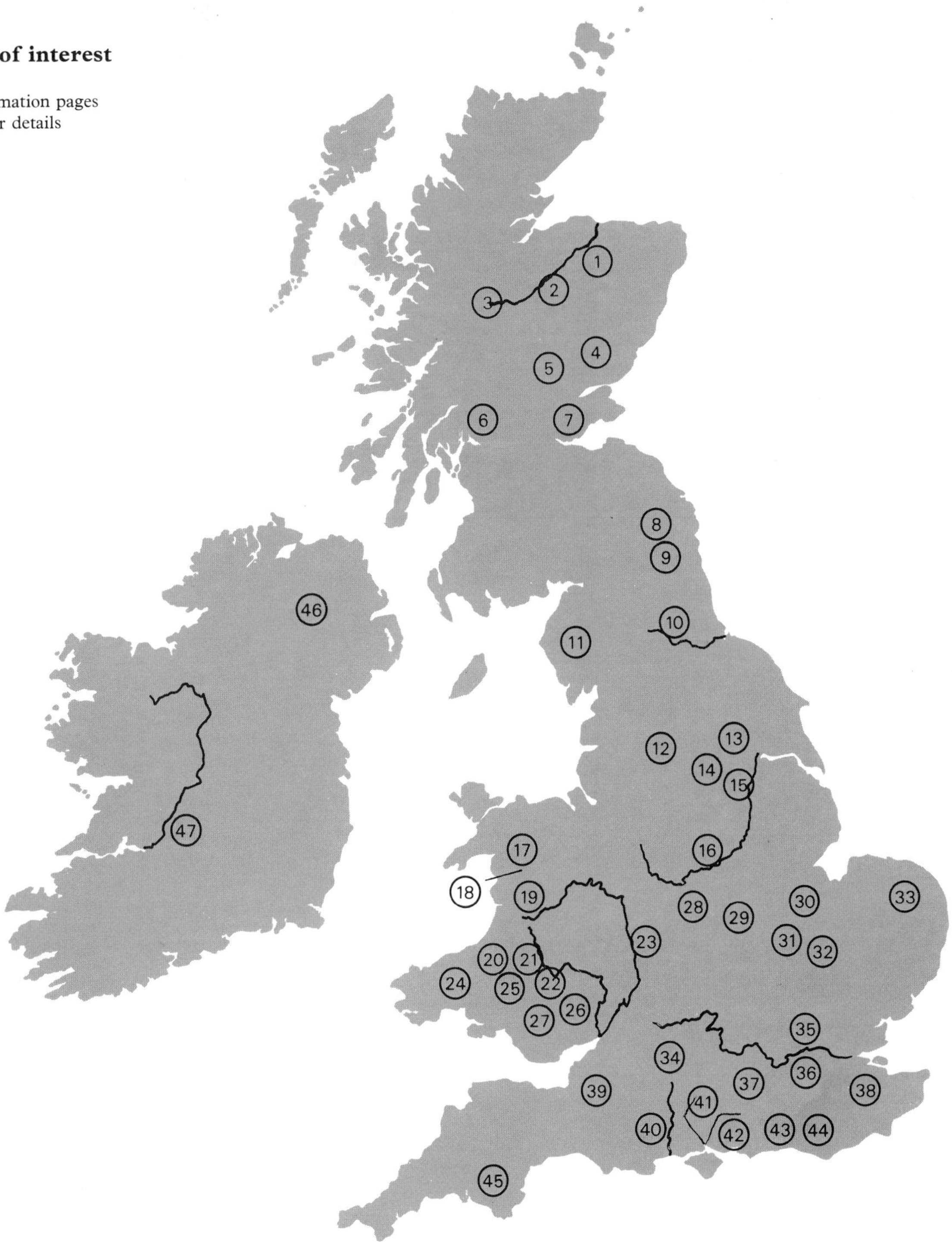

1 Loch Garten, Highlands
2 River Spey, Highlands
3 Loch Ness, Highlands
4 Pitlochry, Perthshire
5 Loch of Lowes, Perth
6 Loch Lomond, Dumbartonshire
7 Loch Leven, Kinross-shire
8 Bolam Lake, Northumberland
9 Shibdon Pond, Durham
10 River Tees
11 Lake District, Cumbria
12 Malham Tarn, Yorks.
13 Fairburn Ings, Yorks.
14 Denaby Ings, Yorks.
15 Potteric Carr, Yorks.
16 River Trent
17 Snowdonia
18 Llygwy Valley, Powys

19 Lake Vyrnwy, Powys
20 Llyn Eiddwen, Dyfed
21 Elan Valley, Powys
22 Wye Valley
23 River Severn
24 Llys-y-fran, Dyfed
25 Brecon Beacons, Powys
26 Brecon and Abergavenny Canal, Powys
27 Tef Fechan, Glamorgan
28 Grand Union Canal, Leics.
29 Rutland Water, Leics.
30 Holme Fen, Cambs.
31 Grafham Water, Cambs.
32 Wicken Fen, Cambs.
33 Norfolk Broads
34 Windrush Valley, Oxfordshire
35 Thames Valley Gravel Pits
36 Thames Valley Reservoirs

37 Wellington Country Park, Hants.
38 Stodmarsh, Kent
39 Somerset Levels
40 River Avon
41 Test Valley, Hants.
42 River Itchen, Hants.
43 Amberley Wild Brookes, Sussex
44 Woods Mill, Sussex
45 Slapton Ley, Devon
46 Lough Neagh, Northern Ireland
47 River Shannon, Ireland

Discovering the Countryside with David Bellamy

Waterside Walks

Country Code

Whenever and wherever you are out walking, please follow these simple rules:

- Guard against risk of fire
- Close all gates behind you, especially those at cattle grids, etc.
- Keep dogs under control
- Keep to paths across farmland – you have no right of way over surrounding land
- Avoid damaging fences, hedges and walls
- Leave no litter – take it away with you
- Safeguard water supplies
- Protect wildlife, plants and trees – do not pick flowers, leave them for others to enjoy
- Drive carefully on country roads
- Respect the life of the countryside – and you will be welcomed.

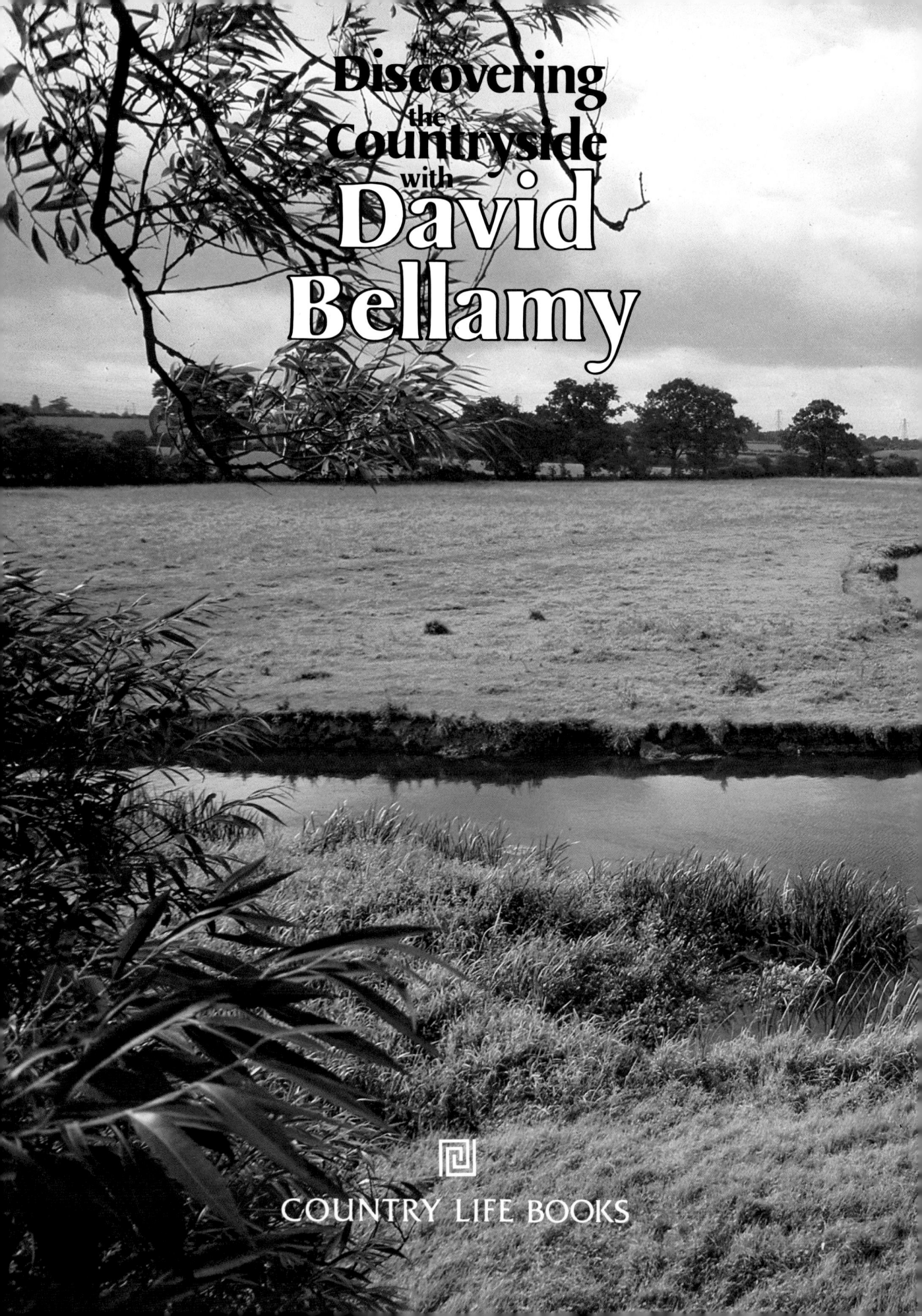

Discovering
the
Countryside
with
David
Bellamy

COUNTRY LIFE BOOKS

Waterside
Walks

Acknowledgements

Line-drawings by Kenneth Oliver

Colour artwork by Keith Linsell

Additional photographs: 25 N.H.P.A. – Ken Preston-Mafham;
63 Wildlife Studies Ltd; 111 Aerofilms;
118 Bruce Coleman – Jane Burton; 119 John Mason
77, 79 Nigel Holmes

The Publishers and David Bellamy would like to
thank the following organisations for their help
in preparing this book:
Royal Society for Nature Conservation
Nature Conservancy Council
Freshwater Biological Association
Sussex Trust for Nature Conservation
Kindrogen Field Studies Centre

In particular we would like to express our gratitude to the
gallant team of experts, Charlie Coleman, Jacqui Morris,
Brian Brookes, Nigel Holmes, Alasdair Berry and Ted Ellis,
whose hospitality, enthusiasm and vast knowledge of the
countryside is only hinted at in these pages.

Published by Country Life Books
an imprint of Newnes Books
84–88, The Centre, Feltham, Middlesex, TW13 4BH, England
and distributed for them by
The Hamlyn Publishing Group Limited
Rushden, Northants, England

© 1983 Newnes Books, a Division of The Hamlyn Publishing
Group Limited

First published 1983

ISBN 0 600 35636 1
Printed in Italy

Photography by Peter Loughran

Foreword

My first intimate contact with mud and water was made in the local brickfields
pond near my home in Cheam, Surrey. There many happy hours were spent
learning the ways of our three species of newt and the insect fauna which
inhabited their waters. My apprenticeship in the watery world was served in the
ponds of Haslemere Educational Museum under the expert guidance of John
Clegg and Arthur Jewell, who showed me the way around both Latin and local
names. The Wandle, Mole and Thames were the first rivers in which I swam
and dabbled a net. Since that time, I have come into intimate contact with rivers
the world over, finding that each, though with a character of its own, can be
understood by the application of the same basic rules. It was Arthur Ransome's
books which lured me first to the English Lakes and then to the Norfolk Broads,
to mess about in boats and look at wetland wildlife and the plants from a new
angle. Since then I have never given a dry look back and waterside walks have
and will always be my favourite walks.

Come splodging with me and the local experts and learn about all the things
which could happen to a raindrop which has fallen onto the diversity of the

Contents

British landscapes. Discover the thrills of life in a mountain stream; the process by which a river passes through youth and adolescence to maturity, as it makes its way to the sea. Explore the beauty and the characteristics of the English Lakes together with the important work carried out by the various research stations of the Freshwater Biological Association. Peer into the clear depths of a chalk stream, and find out what it is like to be a trout or a tadpole; and, finally, marvel at the fact that the Norfolk Broads are man made.

Perhaps, most important of all, relive those days beside the local pond with new meaning and take note that the British landscapes, and that includes our rivers, streams, lakes and ponds, are changing very rapidly, and sadly, mainly for the worse. Once you have read and walked these pages, imagine what it would be like if all the ponds were filled, all our rivers and streams straightened and tamed beyond all recognition, or all our lakes given over to power-boat sports, or – it could happen!

The more waterside walkers who have the wealth of knowledge and experience contained within this book, the more people there will be to speak out against destructive change and for constructive change.

I have only one complaint about this book – it should have been printed on waterproof paper!

A day in the Lakes

Windermere, Derwentwater, Esthwaite are names which conjure up a galaxy of holiday postcard memories from golden daffodils to bursting lungs on the final ascent to the top of Helvellyn. Eutrophic, mesotrophic, oligotrophic are not as emotive but are still words to be conjured with, for each describes a type of lake and from that description you can tell much of what you are likely to encounter on a waterside walk. Come and join me as a raindrop on its way down into Cumbria and become immersed in the subject of lakes and never again look at the contours of a map without planning a waterside walk.

A word of advice, a good map is always useful for a walk in the countryside as not only will it help you find your way there and back but it will tell you a lot about the things you will see en-route and especially about the nature of lakes. One word of caution – water is dangerous stuff and a waterside walk should be planned and undertaken with care.

A magnificent view across Loweswater in the Lake District. These beautiful lakes are the focus of many people's holiday fun and are also one of the best places to explore waterside wildlife in Britain.

Information

My walk takes you through some spectacular lakeland scenery in northwest England, but to appreciate what a marvellous place lakes are you don't have to live near these. All over Britain in recent years lake areas have sprung up creating a new dimension to the natural history of many regions. I am, of course, referring to the gravel pits and reservoirs that supply the sand and ballast for our buildings and roads and the water supplies for our thirsty cities. Nature is always quick to respond to these new environments and given a sympathetic hand can quickly turn a bare, rather unattractive gravel pit into a delightful reed-fringed lake with plenty of wildlife to observe. A classic case is a 90 acre flooded gravel pit in a suburban area in west Kent which within fifteen years turned from a fairly unexceptional pit to one which had over 1300 pairs of fifty-five different species of birds breeding in it, all mainly as a result of a careful management policy carried out by a small but enthusiastic group of naturalists and wildfowlers. These new habitats make excellent places to see wildlife and many that are managed or owned by local trusts have nature trails along their edges. But remember they can be dangerous places as they often have very steep banks: so beware.

Birds

The walk that I will take you on around a few of the Lakes in Cumbria was in high summer, yet I know from visiting these places throughout the year that there are interesting things to see on a walk whatever the time of year. Even a botanist can find a lot to keep him moderately busy in the depths of winter as the mosses and lichens are still there for all to see. But, for the birdwatcher, it is almost a case of the bleaker the weather, the better. Many times I have been out and found huddled groups peering through binoculars and telescopes on exposed vantage points whilst most people would be loath to step out of their front door. So what are they looking at? Well, believe it or not, compared with most of northern and central Europe, Britain gets off rather lightly in winter when it comes to snow and ice, so many thousands of birds regularly book in for a winter holiday on our ice-free inland lakes. These are mostly huge numbers of ducks which have spent the summer breeding out in remote marshy tracts in continental Europe, or have remained in small families and groups hidden amongst dense cover until they all flock together for the winter. A regular check is made on the types of ducks and their numbers so that their populations can be monitored. One of the facts these records have shown is that the vast new areas of artificial lakes are proving to be immensely important for whole populations of these birds.

Red-throated Diver, *Gavia stellata*
Great Crested Grebe, *Podiceps cristatus*
Red-necked Grebe, *Podiceps grisegena*
Slavonian Grebe, *Podiceps auritus*
Black-necked Grebe, *Podiceps nigricollis*
Heron, *Ardea cinerea*
Mute swan, *Cygnus olor*
Whooper swan, *Cygnus cygnus*
Bewick's swan, *Cygnus columbianus*
Canada goose, *Branta canadensis*
Greylag goose, *Anser anser*
Mallard, *Anas platyrhynchos*
Gadwall, *Anas strepera*
Pintail, *Anas acuta*
Wigeon, *Anas penelope*
Teal, *Anas crecca*
Shoveler, *Anas clypeata*
Tufted duck, *Aythya fuligula*
Pochard, *Aythya ferina*
Goldeneye, *Bucephala clangula*
Goosander, *Mergus merganser*
Smew, *Mergus albellus*
Osprey, *Pandion haliaetus*
Coot, *Fulica atra*
Great Black-backed Gull, *Larus marinus*
Lesser Black-backed Gull, *Larus fuscus*
Herring Gull, *Larus argentatus*
Common Gull, *Larus canus*
Black-headed Gull, *Larus ridibundus*
Common Tern, *Sterna hirundo*
Swift, *Apus apus*
Sand Martin, *Riparia riparia*
Swallow, *Hirundo rustica*

Plants

Leaves predominantly submerged
water lobelia, *Lobelia dortmanna*
shoreweed, *Littorella uniflora*
common bladderwort, *Utricularia vulgaris*
spiked water-milfoil, *Myriophyllum spicatum*
mare's-tail, *Hippuris vulgaris*
rigid hornwort, *Ceratophyllum demersum*
Canadian waterweed, *Elodea canadensis*
quillwort, *Isoetes lacustris*
awlwort, *Subularia aquatica*

Leaves submerged or floating on surface
yellow water-lily, *Nuphar lutea*
white water-lily, *Nymphaea alba*
amphibious bistort, *Polygonum amphibium*
pondweeds, *Potamogeton*
starworts, *Callitriche*
duckweeds, *Lemna*

Emergent fringing plants
clubrush or bulrush, *Scirpus lacustris*
reedmaces, *Typha*
yellow flag, *Iris pseudacorus*
bur-reeds, *Sparganium*
common reed, *Phragmites communis*
sedges, *Carex*
rushes, *Juncus*

Sites

The following is a selection of lakes, lochs and loughs with public access to their margins. Many of them are artificially created areas of water either arising as a result of worked-out gravel pits or as reservoirs for the towns and cities. These areas of water are none the less interesting for this and often contain vast numbers of wintering birds. It is suggested that you consult some of the many regional guides, before intending to visit any of the sites, for further information regarding access.

Loch Ness, Inverness-shire. This loch is known the world over as a result of its 'monster'. Frequent searches for this animal have resulted in a vast amount of information being accumulated concerning its wildlife. It is the deepest lake in Britain (130 metres).

Loch Garten, Inverness-shire. A Speyside loch famous for its nesting Ospreys but also contains many natural history features typical of the area. RSPB reserve.

Loch Lomond, Strathclyde. The 'Queen of Scottish Lakes'. This magnificent lake has a deep northern section and a shallow southern section, part of which is a National Nature Reserve. Whole area with the Trossachs forms one of Scotland's National Park Direction Areas.

Loch of Lowes, Perth. A beautiful Scottish loch fringed with woodland. Ospreys breed here and there are also many species of waterfowl in winter. Good also for water plants such as water lobelia. Scottish Wildlife Trust reserve with visitor centre.

Loch Leven, Kinross-shire. A shallow productive loch in Central Scotland. An important wintering area for a large number of waterfowl. Part National Nature Reserve, part RSPB reserve.

Bolam Lake, Northumberland. A country park with lake and woodland areas including nature trails and an information centre. The nearby Kielder reservoir was opened recently and has an information centre.

Lake District, Cumbria. The beauty and interest of the Lakes is well-known. This remarkable area has a great range of lake types with Wastwater and Ennerdale showing highly oligotrophic features and the more developed lakes such as Windermere and Esthwaite showing eutrophic features. Smaller corries and lakes, such as Angle and Stickle Tarns can be found high up in the hills.

Malham Tarn, Yorkshire. An upland tarn with interesting geological associations. Partly owned by National Trust. This lake is the site of one of the Field Studies Centres which has an information centre and runs courses on various aspects of natural history.

Lake Vyrnwy, Powys. A nineteenth century reservoir built high up in the Welsh mountains. Many interesting upland and water birds including dippers. Nature trails. Spectacular scenery. Part RSPB reserve.

Llyn Eiddwen, Dyfed. An upland oligotropic lake with interesting wintering wildfowl and water plants such as quillwort and water lobelia. West Wales Naturalists' Trust Reserve. Access only along west side of the lake.

Llys-y-fran, Dyfed. A country park in the Preseli Mountains based around a large reservoir.

Lough Neagh, Northern Ireland. The largest freshwater lake in the British Isles (over 400 square kilometres). There are extensive areas of reedswamp and fen along the lake margins. Nature trails can be found at Shane's Castle Reserve (RSPB) and Oxford Island. The whole area is a wildfowl refuge of international importance.

Rutland Water, Leicestershire. A huge reservoir which is an important area for waterfowl, and waders. Part of the water has public access and part is managed as a reserve by the Leicestershire and Rutland Trust for Nature Conservation.

Grafham Water, Cambridgeshire. A man-made reservoir which provides good opportunities to watch waterfowl with public footpaths around many stretches of the lake margin. Part of the reservoir and its margin is managed as a reserve by the Beds. and Hunts. Naturalists' Trust. Nature trails.

Thames Valley gravel pits. Extensive gravel pits have been excavated in the areas around London to provide building materials, etc. These pits have often been allowed to become flooded to form large areas of open water, many of which have been colonized by plants and animals, particularly water birds, which use them for roosting areas in the winter. Some have been leased or managed by local conservation groups or authorities and have nature trails. An example of this are the Broad Colney Lakes, Colney which are leased by the Herts. and Middlesex Trust for Nature Conservation.

Thames Valley reservoirs. The creation of large reservoirs, particularly in west and south London has provided an important new habitat for waterfowl. Special permits are required to visit most of these. However, some have limited public access and provide superb birdwatching sites. The public causeway that crosses Staines Reservoir, Middlesex, is generally good for waterfowl, especially during the winter months.

Wellington Country Park, Hampshire. A country park centred around a series of flooded gravel pits, which have interesting wintering waterfowl. Has 'leisure' areas as well as nature trails and woodland walks.

A day in the Lakes

If you want to visit an area with plenty of waterside walks then one of the best places to go is the Lake District. So here I am, standing beside a beautiful lake on one of those typically rainy days in the British countryside – it is not only pouring but it also happens to be blowing so hard that the rain is hitting us horizontally. But what better time to actually explore the importance of water in the landscape? The weather doesn't really matter, as you can admire the view from the comfort of your car. But I never do that, as I like to get stuck in and have a good paddle around along the lake-side where I get wet from the top and the bottom at the same time until I almost become part of the lake itself. However, if we are going to do this we ought to understand what lakes are all about.

First of all, I can hear you say: 'What are all these lovely stretches of water doing here in the rainy old Lake District?' Well, it's partly to do with the rain, but in order to understand their origin you have to think back to 10000 years ago to the time of the Ice Ages. A great ice sheet once covered the land hereabouts, and the centre from which most of the ice radiated in this part of the world was in the high mountains of the Lake District. Gradually as the temperature rose the ice sheet became a series of massive glaciers which gouged out deep U-shaped valleys between the mountain tops. Then as the temperatures rose still further the glaciers themselves began to melt and retreat back up the valleys, depositing the material that they had picked up on the way down. Sometimes this material formed an effective barrier across the bottom of the valley. As the meltwaters from the snow and ice came tumbling down the mountains and valleys they found themselves in a cul-de-sac and slowly began to fill up the valley bottom, forming the Lakes that we can see today. Obviously there is a lot more to it than that, but all you have to do is to look at a map of the area and you can see immediately that the shapes of the lakes closely follow the great fingers of the glaciated valleys that radiate out from the centre of Cumbria. These lakes can also be very deep. For instance, Windermere is over sixteen kilometres (ten miles) long and at the deepest point is sixty-six metres (220 feet) deep, yet is only 800 metres (880 yards) wide.

Now, I reckon that if I had a chance to be something else, I'd like to be a raindrop. They travel this world for free. They don't have to buy air tickets or motor cars. They

When the rain is pouring down you can always admire the view from the car but I love nothing better than getting out there and soaking up the Lakeland scene! Here I am standing on a wave-cut platform in Crummock Water.

get evaporated up into the clouds and then off they go. It must be a blooming exciting thing to be a raindrop coming along from the west, not knowing where you are going to fall. But you can usually be safe in assuming that quite a lot of it is going to fall on the hillsides of the Lake District. Here I am imagining that I'm a raindrop and I'm about to come down on a ridge of land that separates two lakes, one called Loweswater and the other is called Crummock Water. Now, if I fall on one side of the ridge I am going to land on some nice gentle slopes amidst fairly rich pastures grazed by sheep and cattle. If I fall on the other side I'm going to land on a bleak slope with little soil cover. As I make my way down to the lake I'm going to pick up all sorts of minute particles from the rocks and the soil, and these are going to be washed along with me into the lake. So, you see, the nature of the surrounding land is going to have an important effect on the lake water itself.

If we look around Crummock Water, the amount of land that is accessible for reasonably productive farming is very low. A great deal of it looks fairly rocky and acid. This means that our vast armies of raindrops will

pick up very few mineral nutrients to feed into the lake. The lake can, therefore, be classed generally as unproductive, which means that few or no plants and animals will be found in it. We can test this by looking along the shore line to see what plant and animal remains have been washed up. There is a limited strand line here with a few torn up plants and twigs. But I can't see any nice banks of waterside plants. What we have here, though, illustrates another important factor – wind action.

As I have said it is enormously windy today, in fact, it feels as if it is gusting up to a force nine gale at times. Looking out from where I'm standing (only just!) I can see waves over a foot high. I could almost be forgiven for thinking that I was standing on the beach at Brighton on a blustery day. All this wave action is going to eat away at the exposed edges of the lake, forming a series of shingly or rocky platforms that will be very inhospitable for plant growth. In sheltered bays where you don't have this wave action you have a build up of organic debris which has been slowly deposited, and this provides an ideal site for plants to colonise. However, in an unproductive lake, this process will take a very long time. These lakes are called 'oligotrophic' which is a term that is derived from the Greek: *trophe*, food and *oligos*, little or few, as opposed to a eutrophic lake (Greek *eu*, well), which we will look at next.

Here we are only a few miles further along the road and I am currently walking across a farmer's field which looks as if it has been reseeded and fertilisers have almost certainly been put down. In front of me is a great flock of sheep and up on the hill is a small herd of cows. If I keep to the public footpath and remember to close the gates after me then I shall do no harm. Our destination is the section of path that runs along by the far side of Loweswater. Already, just from looking at this surrounding farmland, I know that we shall find a very different situation by the lake-side.

We have walked around and here is a channel that the farmer has dug to help drain his land and, who knows, it might just be carrying our raindrop. Unlike the other side of the ridge, this part of the valley has a much gentler slope and a reasonable covering of soil. So it is likely that the rainwater draining off this slope will have picked up a fair amount of mineral nutrients such as phosphates, nitrates and potassium.

This particular side of Loweswater is also quite sheltered and if we look at the shore line we can see that it has a lot of silt deposited at the edges. However, even though the water is a good deal calmer then Crummock, the water is not that clear. This is mainly due to silt carried in suspension in the water from the stream. Earlier on in the year we would have been able to see another important factor which would affect the clarity of these eutrophic lakes – algal blooms.

Although these lakes are very deep during the summer, they tend to have two distinct temperature layers: a colder more dense layer in the deeper reaches known as the hypolimnion, and a noticeably warmer layer on top of this, the epilimnion. During the winter when the lake cools down, these layers become mixed. But in the spring the upper reaches are increasingly warmed by the sun, eventually forming a distinct layer that does not mix with the lower one. It is in this upper warm layer that most of the 'production' in the lake occurs. In particular, algae, such as *Asterionella*, a minute diatom, increase in large numbers in the spring sunshine. It is this floating mass of miniature plant life that forms the basis of the food chains in the lake,

so without them and their ability to turn the dissolved mineral nutrients into living matter the lakes would be almost dead. However, as with all things, a balance is needed. If the lake becomes too eutrophic, as a result of the farmers spraying their fields with increasingly large amounts of fertilisers or villages and towns pumping out uncontrolled amounts of effluent, the lake will soon start to suffer, as not only will the light penetration in the water be impaired but also the increasing amount of oxygen that the algae consume at night and during their decay will mean that dangerously anaerobic conditions, especially in late summer, can be produced where neither animals or large plants will be able to survive.

I like to think of the microscopic algae and the tiny animal plankton feeding on them as a very thin animal and vegetable soup. But when the soup becomes quite thick, how do the larger macrophytic plants survive as the life-giving rays of the sun will only be able to penetrate a little way beneath the surface? The answer, of course, is that they root in the rich silty bottom of the lake and stick their leaves out either on the surface of the lake or

A bubbling stream carrying rainwater that has fallen on the surrounding slopes down to the lake. This water will have picked up mineral nutrients, which will be washed into the lake contributing to its overall productivity.

Left **The flowering panicles of a rush. These plants have evolved flowers that are wind pollinated and the petals therefore have no need to be attractive and have as a consequence been reduced to brown scales.**

above it. If we look out across the water here, at a depth of between two and three metres, we can see the first colonisers of the lake edge doing precisely this. They are the large flat leaves of both species of water lilies that we have in Britain: the white water lily, *Nymphaea alba*, and the yellow water lily, *Nuphar lutea*. I prefer its other common name – brandy bottle – as the flowers don't last long and they then produce a wonderful fruit capsule that looks like one of the old brandy bottles. Down underneath the floating leaves are others that look partly like cooked cabbage leaves; it is quite usual for an aquatic plant to have two types of leaf.

The floating lily pads can be a problem as they can quickly cover the surface of a small pond or sheltered bay, effectively blocking out the light for everything else. However, our next zone of plants has learnt to cope with this. These larger water plants are not eaten by many of the lake animals and so they often leave a considerable amount of decaying vegetation when they die back in the winter. This all helps towards the build up of a rich organic bed of silt on the lake shelf here. As we have commented, if you live on the edge of a eutrophic lake the sensible thing to do is to grow up and out above the water so you can get to the light. This is what the next colonisers do, the perennials. These are usually large plants such as the common reed, *Phragmites*. Our chief emersive perennial here is the bulrush or common clubrush, *Scirpus lacustris*. These plants have formed a dense stand which has resulted in a further

Opposite **As soon as the Lakes were formed nature began slowly to claim them back. If one looks along the edges of many of the more sheltered Lakes one can see this process of gradual colonisation. This lovely view shows two of the initial plants that help to bind the silt and create an environment where other plants can move in. In the foreground are some rather battered lily pads of** *Nymphaea alba*, **the white water-lily and behind these is a superb stand of the bulrush or clubrush,** *Scirpus lacustris*.

Behind the stand of bulrush, the water is shallower and has been colonised by a community dominated by bottle sedge, *Carex rostrata*. The vegetation here is kept reasonably open by the grazing cattle.

accumulation of silt and organic debris, effectively pushing out the water lilies. All this is gradually raising the level of the bottom of the lake in the area. So we are looking at a process which is causing the lake to shrink. The term for this plant succession is hydroseral development. This gradual silting up and colonisation of the lake will have been continuing since the lakes were first formed over 10000 years ago. Eventually, if the process continues, the entire lake will become a swampy marsh and begin to dry until it is finally taken over by trees forming a woodland which in this part of the world is the climax vegetation.

That's the theory. Now let's get out there and have a look at the plants themselves. Ooh, it's really very cold and squelchy. I think I'll just go out as far as the start of the bulrushes. Well, here I am, the water is up to my thighs and it's blooming fantastic. It's started to rain again but I'm enjoying myself as I have found some super things to tell you about. First of all, now that I have got my head in amongst the bulrushes, I can say that they only cover about a third of the surface of the water but as they are so tall they have succeeded in shading out all the other plants. They must be over a metre high as they are above me, which means that there is more than a metre growing under the surface.

As I splodge my way back to dry land I have now come to a transitional zone

between the reedswamp community with the tall bulrushes and a much more diverse community which we might call sedge fen or even sedge meadow. It is quite an abrupt change, for within a metre I have moved from a solid stand of one species to a rich mixture of bottle sedge, *Carex rostrata*, water horsetail, *Equisetum fluviatile*, and a variety of other associated plants. The few bulrushes that are growing amongst the sedges are much shorter and don't look nearly as healthy as their cousins further out. So here we can see the successional change at work. The build up of organic matter and silt which the rushes have helped to create has resulted in an area of shallower water which has favoured the sedges. Another factor is that this is probably as far out as the grazing cattle dare go. Their frequent forays out here to nibble away on the lush plant growth has also given the shorter growing sedges an advantage. The bottle sedge has this dense cylindrical mass of fruits which distribute its seeds of success. It is a perennial and uses its submerged network of rhizomes to store up food during the winter months, ready to push up masses of new shoots every spring. These weird looking plants like mini-Christmas trees are horsetails. They are the modern descendants of the giant horsetails that formed a major part of great swampy forests during the Carboniferous Period over 300 million years ago.

If I look down in between all the sedges I can see quite a few other plants. Here we have the round leaves of marsh pennywort, *Hydrocotyle vulgaris*, which, believe it or not, is a relative of the tall parsleys of the hedgerows. Its leaves look just like tiny umbrellas and if I was a gnome I would pick one up and use it to shelter from all this rain. That's lesser spearwort, *Ranunculus flammula*, a marsh-loving relative of our garden buttercup. Now, if I want to really get to grips with some of these plants I'm going to have to grovel around amongst the bases of these plants with my hands. It's very peaty and the smell that is coming off is pretty fantastic. It's like bad eggs which tells my nose that it is hydrogen sulphide, confirming that there is a lot of decaying organic matter down there. Here we have a nice little rush, *Juncus bulbosus*. This plant either forms little tufts or simply floats between the other plants. You can always tell that it is bulbous rush by feeling the base of the plant, where you will find that it has a mass of little bulbils.

Above the smell of the hydrogen sulphide I can catch something that is altogether much more pleasant – the aroma of water mint, *Mentha aquatica*. We are now in much shallower water and I can stick my hand into the ooze in the bottom and see what the floor is like. It looks just like a mass of brown decaying matter, which is what you would expect. The stringy bits are the rhizomes of

the bottle sedge, but if I rub some of the deeper material between my fingers I can feel some sharp, hard bits which are particles of mineral matter that have been deposited here. So we can say that one of the reasons for the success of the hydrosere just in this bay is the mineral enrichment of the lake deposits, probably coming in from that nearby stream.

I have now moved out of the marshy fen community and I'm standing on dry ground again. We have a rather interesting situation here as the farmer has put up a barbed wire fence stretching down the hill right out into the lake and stopping in the stand of bul-rushes which, as we have already guessed, is as far out as the cattle are likely to venture. This side of the fence has been regularly grazed and the sedge-rush community is still dominant in the wet meadowland. But on the protected side we have a woodland that extends almost down to the shore of the lake. It is predominantly made up of willows, *Salix*, which don't mind getting their feet wet. This type of woodland is the natural successor to the transitional sedge fen com-munity. If we look further up the slope where the accumulation of leaf litter and fallen twigs has raised the level of the land still further, we can see birches, *Betula*, growing. These in turn can provide a 'nurse-crop' for oaks, *Quercus,* to become established. So there it is – an almost perfect example of a hydrosere, something which is going on

Above left **Growing amongst the peat-rich bottom of the marsh, is a relative of the garden buttercups, lesser spearwort,** *Ranunculus flammula.* **They are named after their pointed leaves and are poisonous.**

Above **The distinctive circular leaves of marsh pennywort,** *Hydrocotyle vulgaris,* **amongst a rich mixture of mosses, rushes, sedges and the toothed pinnate leaves of marsh cinquefoil,** *Potentilla palustris.*

Right **Where the natural succession or hydrosere has not been altered by man, woodland, which is the climax vegetation, will develop on the higher land as shown here, only metres further along the shore from our sedge meadow. These are hardy willow trees,** *Salix*, **which, like me, don't mind getting their feet wet.**

wherever there is a build up of silts which can allow the plants to start colonising the fringes of these superb lakes. Now I want to show you something very different – a very oligotrophic lake, and we'll see what goes on there.

After having a gentle stroll around the farmland at Loweswater we are now looking at a complete contrast – dramatic fells rising almost sheer out of the waters at Ennerdale. The tops of the scree-covered slopes are hidden in the clouds and a strong wind is gusting across the lake surface. Any rain that is going to fall in the catchment of this lake is going to tumble and cascade down the steep slope, forming little waterfalls and swift-flowing becks. Looking at the large amount of bare rock and the thin covering of acid grassland, this rainwater is going to pick up very few mineral nutrients. This is a pretty sure sign that the lake water itself is going to be very unproductive – a highly oligotrophic lake, in other words. The result of this lack of nutrient input into the lake will be that there will be little phytoplankton. An easy way of assessing this, as we noticed in our previous lake, is the clarity of the water. So let's make our way over and see what it's like.

Here I am standing on a boulder by the lake side and sure enough the water is very clear. I can see lots of different coloured bits of fragmented rock that have probably come down from the scree slopes; they look like hard volcanic rocks. There is very little sediment and if I look behind me there is not much other than scraggy bits of acid moorland and scree, quite unlike our eutrophic farmland. Unfortunately the lake bottom shelves away quite steeply just here so we will

Opposite **Having had a gentle stroll around Loweswater, the almost sheer fells surrounding Ennerdale make a dramatic contrast. Any rainwater running off these slopes is going to contribute very few nutrients to the lake water, therefore it is not surprising to find that the lake is very oligotrophic.**

have to go around to another part of the lake to look at its specialised plants.

Well, we have found a section with a nice gentle shelf to it and I am actually up to my waist in this nice clear water. Looking down I can see my tatty plimsoles, which shows that it is nice and clear! Now, if I was making a proper study of the lake I would be out in a boat armed with a Secchi's disc, which is a round white board that you let down into the water to the depth where it is just visible. The depth at that point is read off and this is then used as a measurement of the light penetration in the lake. In Ennerdale it is just over eight metres (26 feet), which means that the lake is very oligotrophic. Anyway, at the depth of half a Bellamy, just here I can see a number of plants growing on the floor of the lake. They are pretty widely scattered but I can make out *Littorella uniflora*, shoreweed. That is a good name as 'littoral' means the land between the tides, and although lakes don't have tides at least it tells me that I'm dealing with a plant that grows in water. This relative of the familiar plantains is an interesting plant as it only flowers when the lake level goes down in the summer, and when it is just emergent it puts out tiny white flowers. Here is another plant with tufts of green needle-like leaves – quillwort, *Isoetes lacustris*. As I walk towards the shore the two plants are becoming more dense, almost forming a squashy green carpet. This is because there is more sunlight reaching them for photosynthesis. Over here is another underwater plant forming quite large patches all of its own. It is water lobelia, *Lobelia dortmanna*.

Now, you needn't do a Bellamy and wade about in the water as you can often find these plants washed up on the lake shore. They all look more or less the same at first glance, but a closer look will enable you to identify them easily. The ones along the side here are already dead so it won't do any harm if I break a leaf off and look inside. A hand lens would help here, but if your eyes are reasonably all right you should be able to see enough to tell them apart. If the leaf is pointed and fairly tough and is made up inside of a solid spongy tissue, then you are dealing with *Littorella*. If it is stiff but when you break it open you can see that it is made up of four tubes which run down the whole length of the leaf, then you have *Isoetes*. This is a very interesting plant as it is a relative of the ferns. So whatever time of the year you find it there will be no flowers. But if you feel the base very gently you might find that it is all lumpy, which will be the sporangia. Don't muddle it up with bulbous rush which is a much more delicate looking plant and will almost certainly have flowers or fruit. Now, how about water lobelia? The rosettes of leaves of this plant are quite soft and fleshy to touch and each leaf is arched back at its tip. If you are still not sure, you can break open a leaf and if you look inside you should find two whacking great tubes. If you are lucky, you might come across it whilst it is in flower and then you can see its lovely lilac-pink petals blowing in the wind above the surface of the water. It is one of my favourite plants and very, very common in certain oligotrophic lakes in Scotland.

I am now going off to look for another bay where perhaps there has been a little deposition along the shore, to see if we can find some emergent plants. A close look at a good map of the area will always help.

I have now arrived at a little sheltered bay and in amongst the stones I can see some sediment, although there is no obvious succession of plants as at the eutrophic Loweswater. Right along the edge of the lake I have found a typical plant of boggy land and more acidic waters – Bogbean, *Menyanthes trifoliata*. This is one of the easier waterside plants to identify as it has a three-part leaf, rather like a gigantic clover leaf. Earlier on in the year it has lovely pink and white flowers with delicate fringes to the petals. If you find one washed up along the lake side have a look at the stem. You will find that inside they are made up of a wonderful squashy tissue. This tissue is called aeren-

The flowers of bogbean, *Menyanthes trifoliata,* **showing their beautifully fringed petals, almost like a piece of Turkish towelling.**

chyma. This consists of living cells with whacking great holes in between, rather like a string vest. All the water plants of the world except those that live in highly oxygenated waters have this type of tissue, which not only allows passages for oxygen to circulate from the aerial sections down to the roots but it also provides support with the minimum number of cells. I always think it interesting that if you were to look at the inner skin of a modern aeroplane you would find that its structure was remarkably similar to the spongy aerenchyma of a water plant, which, of course, the plant kingdom came up with millions of years before man.

If we look at the shore here we can see that instead of graduating into farmland we have a bog moss, which is very acid and poor in nutrients. It is made up of my favourite plant – *Sphagnum*. But that will be the subject of another walk, in another book.

So we have seen two extremes in Loweswater and Ennerdale and I am sure you will find lots of variations and combinations of the two types that we have looked at today. A close look at an Ordnance Survey map can often tell you what sort of lake you are likely to find. Gentle slopes with woodland or farmland on the lake shore is usually an indication of a productive landscape and, therefore, a eutrophic lake, whereas steep, rocky slopes with short becks and little surrounding woodland could mean that you have an oligotrophic lake. Perhaps you could make a survey of the lakes of your area and put them into some type of order as they have done here in the Lake District. Another important thing to do when looking at lakes is to visit them at different times of the year, so that when you have all your notes together you can get a complete picture of the place. Right now I don't think there is one bit of me that isn't wet through, so I'm off home to dry out and have a pot of tea before I write up my note book.

Ponds in summer

Tadpoles in a jam jar and the squelch of mud and duckweed between your toes, that's what ponds are all about. Well, that, and a lot more, because, in terms of seeing a cross-section of plant and animal life, a healthy pond is almost as good as having a coral reef right in your own neighbourhood.

Village ponds were once an important part of rural life – a swimming place for ducks and geese and a watering place for farm stock and fire brigades in times of drought and emergency. So they were regularly dredged, cleaned and kept in good, watery heart. Likewise, many of man's excavating activities left subsidence areas and all other sorts of water-filled holes, which soon became refuges for the local aquatic flora and fauna. Unfortunately, what man can create he can also destroy and recent years have seen the destruction of many of our ponds. Those which are left are thus of increasing importance to nature conservation, amenity and education.

If you have a local pond learn all about it and be proud of what you have.

A coot with her chicks searching for food amongst a lily-covered pond. When one sees a picture like this one wonders whether the expression 'bald as a coot' refers to the white plate on the forehead of the adult or the lovely chick with its thinly covered head and mutton-chop whiskers!

Information

Our walk looks at pond life in mid-summer when the vegetation is at its thickest and the most eye-catching of the animals are usually the insects darting across the surface and busily feeding on the fringing plants. But if you visit your local pond much earlier in the year you might be lucky enough to either discover some clumps of frog spawn or see the wriggling tadpoles swimming in the shallower water at the edge of the pond.

Amphibians

There are only six native species of amphibian in Britain, consisting of three species of newt: the common newt, *Triturus vulgaris,* crested newt, *T. cristatus,* and palmate newt, *T. helveticus;* two species of toad: the common toad, *Bufo bufo,* and natterjack toad, *Bufo calamita*; finally, there is the familiar common frog, *Rana temporaria.* To this list must be added the introduced marsh frog, *Rana ridibunda,* which has to some extent replaced the common frog around Romney Marsh in Kent.

Newts lead a terrestrial life for most of the year, spending the day hidden amongst rocks and logs, feeding on worms and insects at night. The best time to see them, however, is when they return to the water to breed, usually in March. Like frogs, they prefer ponds and slow-flowing rivers and canals. During this time they lead a wholly aquatic life, feeding on aquatic invertebrates. They all have impressive courting rituals where the males rub the snout of the females; at this time of year the coloration of the males becomes increasingly intense and a crest also develops. The male palmate newt develops webbed feet as well. Unlike frogs and toads, the eggs are deposited singly on submerged vegetation. The tadpoles remain in the water until late summer when they leave to take up a more terrestrial existence. As with other British amphibians, they hibernate through the winter.

Frogs and toads

There was a time when it seemed as if nearly every ditch and pond had its collection of early spring frogspawn. Sadly, today, the common frog is not as common as it once was. It has been said that the mass collection of the frogspawn for junior school nature lessons contributed to this decline. However, although it is obvious that this did not help and should be carried out with more consideration today, loss of habitat is probably a far more important factor. With the general provision of piped water for feeding livestock, the farmer has no longer any need to maintain the ponds in his fields, and the treatment of dykes with insecticides and the improvement in land drainage has also resulted in a dramatic decrease in suitable breeding waters. In some areas, garden and ornamental ponds now constitute the major habitat for this delightful animal. These ponds need not be large to attract frogs and provision of such a pond is something which any garden-owner can do.

Toads will also use garden ponds and as a bonus will feed on the slugs in the vegetable patch! Their spawn, unlike the dense masses of the frog, is produced in long ribbons which the female toad wraps around aquatic vegetation. The adult toads are more able to tolerate dry conditions and can be told from frogs by their shorter hind legs and warty skin. Both the toad and the frog have favourite breeding ponds and can sometimes be seen migrating in large numbers to these in early spring. Some authorities even have signs up to warn motorists, where the migration route crosses a road!

The very rare natterjack toad is a creature of sandy places and is now found in only a few places in the south of England and an area of sand dunes in the North West. It is so rare that it is legally protected under the Wildlife and Countryside Act.

Plants and animals to look out for:

Plants

Submerged or partially submerged plants in water:
yellow water-lily, *Nuphar lutea*
white water-lily, *Nymphaea alba*
pondweeds, *Potamogeton*
duckweeds, *Lemna*
starworts, *Callitriche*
spiked water-milfoil, *Myriophyllum spicatum*
amphibious bistort, *Polygonum bistorta*
frogbit, *Hydrocharis morsus-ranae*

Fringing plants:
arrowhead, *Sagittaria sagittifolia*
water plantain, *Alisma plantago-aquatica*
bogbean, *Menyanthes trifoliata*
reedmace or bulrush, *Typha*
lesser pond sedge, *Carex acutiformis*
rushes, *Juncus*
common reed, *Phragmites communis*
yellow flag, *Iris pseudacorus*
sweet flag, *Acorus calamus*
reed sweet-grass, *Glyceria maxima*
bur-reeds, *Sparganium*
flowering rush, *Butomus umbellatus*
water forget-me-not, *Myosotis scorpioides*
water figwort, *Scrophularia aquatica*
water speedwell, *Veronica anagallis-aquatica*
hemp agrimony, *Eupatorium cannabinum*
great willowherb, *Epilobium hirsutum*

Aquatic invertebrates to look out for:
sponges Porifera
pond sponge, *Spongilla lacustris*
Flatworms Platyhelminthes
    Polycelis nigra
Leeches Annelida
    Melobdella stagnalis
Molluscs Mollusca
ramshorn, *Planorbis planorbis*
great pondsnail, *Lymnaea stagnalis*
wandering snail, *Lymnaea pereger*
horny orb-shell, *Sphaerium corneum*
Arthropods Arthropoda
    Cyclops
water louse, *Asellus aquaticus*
freshwater shrimp, *Gammarus*
fish louse, *Argulus foliaceus*
pond olive mayfly, *Cloeon dipterum*
large red damselfly, *Pyrrhosoma nymphula*
common ischnura, *Ischnura elegans*
common sympetrum, *Sympetrum striolatum*
common aeshna, *Aeshna juncea*
water cricket, *Velia caprai*
pond skater, *Gerris*
water scorpion, *Nepa cinerea*
backswimmer, *Notonecta*
lesser water-boatman, *Corixa*
brown china mark moth,
    Nymphula nymphaeta
cinnamon sedge caddis fly,
    Limnephilus lunatus
mosquito, *Culex*
midge, *Tanypus*
whirligig beetle, *Gyrinus natater*
great diving beetle, *Dytiscus marginalis*
great silver beetle, *Hydrophilus piceus*
water spider, *Argyroneta aquatica*
water mite, *Hydrodroma despiciens*

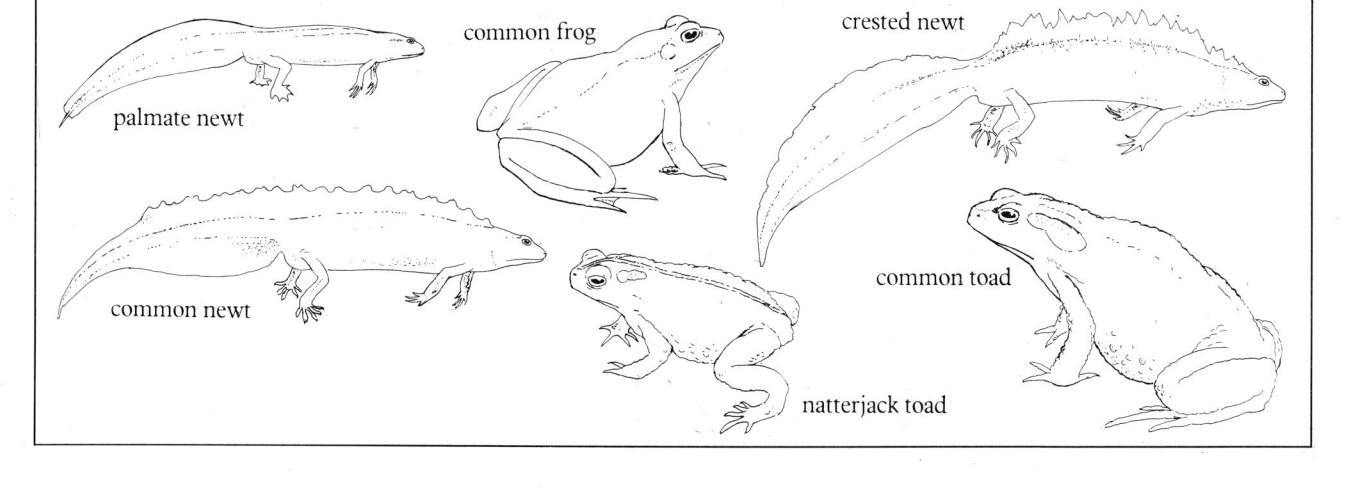

palmate newt

common frog

crested newt

common newt

common toad

natterjack toad

Ponds in summer

with
Jacqui Morris
and
Charlie Coleman

The large pond or 'lake' at Woods Mill, carpeted with yellow water-lilies, *Nuphar lutea*, **and surrounded by a fringe of rushes, flags and reeds. These unpolluted ponds are an ideal habitat for a host of birds and insects and always provide interest no matter what time of year.**

Ponds are among the most accessible of the freshwater habitats available to us in Britain. Who can resist dabbling in a pond on a warm summer's day and falling under the spell of its surroundings and the immense variety of living things that are dependent on it? Ponds also represent water in a calm, subdued form on a human scale and may therefore be the perfect place to introduce children to the wealth of natural history that surrounds them.

Where better to begin this story than at Woods Mill Nature Reserve near Henfield in Sussex. Woods Mill, the Headquarters of the Sussex Trust for Nature Conservation, has been described as 'an extraordinary piece of countryside' with ponds and woods surrounding its eighteenth century mill. On a sunny June day we went down to its lush grounds to look at the ponds which, for the most part, have been scooped out of the soft wealden clays by man. Beyond Woods Mill the ground starts to rise towards the chalk escarpments of the South Downs where the springs which feed the ponds can be found. Our guides were Jacqui Morris, former Education Officer with the Sussex Trust, and Charlie Coleman, the Warden of Woods Mill.

Just outside the mill we first stopped by the stream which feeds the great mill wheel to admire a pair of nesting grey wagtails,

Motacilla cinerea, with their bright sulphur-coloured under-bellies. We then slowly walked on through the reserve to a large pond edged with a thick ribbon of reeds and other plants. Our first impression was that it was teeming with life, a view which was soon confirmed.

However, we started by asking Jacqui about what to look out for if you want to find a good pond to explore.

'A diversity of plant growth would be the first thing to search out. A pond should have a really healthy look about it with a good range of plant life. This provides the basis of the food chain within the pond and therefore it supports the wealth of animal life we all like to see. On the other hand, if you have a pond whose surface is completely covered with duckweeds or algae, no light, which is essential for plant growth, will penetrate, so that water will probably not be able to support much else.

'Obviously the pond should be clean and unpolluted. The presence of dragonflies and damselflies indicates a healthy pond. A good pond to study would be one with a fair degree of stability, so you can study it throughout the year. Impermanent ponds shouldn't be overlooked, though, because they have their own specialised animals and plants. This pond is pleasantly surrounded by trees, yet they do not overhang it and keep the sun out. This can be a problem with a smaller pond. Living things need sunlight and an entirely shaded pond will be almost devoid of life. There is a pond in my home village which is shaded in the main by beech trees. Years ago fish were put in, but they've never bred. They've grown to a fair size and have "mud-spots", the little fungal growths along their sides but there are never any fish fry. And no pond weeds at all. It is a good example of what can happen when a pond is too densely shaded.'

Charlie commented that the large pond we were looking at was almost entirely artificial.

'It was, for the most part, dug out by an owner in the 1940s. The area was marshy and, in fact, we have left a small swampy area at the far end of the pond. It would probably all silt up if we didn't remove some of the water lilies in the winter-time.'

Jacqui went on to explain what would happen to this pond if it were ever allowed to silt up.

'A pond is a transient habitat and if it doesn't have a steady water source or if it is allowed to silt up reeds and sedges would

The striking flowers of great willowherb, *Epilobium hirsutum*, showing the distinctive stigma with its four creamy lobes. A local name for this large waterside plant is codlins and cream.

gradually fill it in. The pond would become marshy and tussocky. Then alders and other trees would take root on the tussocks. Over a number of years the area would become woodland and eventually it would revert to oak woodland, which it was centuries ago, as was most of the south of England.

'To get back to our pond here at Woods Mill, this is a good example of what can be done to create an attractive pond that is sympathetic to its surroundings and which can be maintained for the benefit of wildlife and the enjoyment of visitors. I want to emphasize that garden ponds and other types of artificial water are just as important as natural ponds. They too provide an important habitat for a very large number of invertebrates and plants that depend on still water. Many insects, for example, need an aquatic environment in order to complete the early stages of their life cycles. These types of pond are more important than ever before because so many ponds have disappeared and are continuing to disappear. Farm ponds have been filled in, as have village ponds. Dew ponds, an ancient and particularly vulnerable form of pond, have been polluted, have silted up or have been filled in.'

As we walked beside the pond we noticed quite a lot of bird life on the open water and commented on some brown, rather drab-looking ducks.

'They are mallards and they are loosing their breeding plumage just now. The drakes especially are beginning to look a bit shabby because they are coming into "eclipse" plumage which probably enables them to remain unnoticed while they are moulting their flight feathers. As you can see, they now closely resemble the females. You can always tell the male, though, by his greenish-yellow beak.'

Around the edge of the pond there were shoulder-high ranks of sedges and reeds. Among these were some stands of sweet flag, *Acorus calamus*, and reedmace, *Typha latifolia*.

'Sweet flag, which tends to grow in a mass, has curious horn-shaped flowers. The leaves are crinkled along the edges and give off a delightful sweet smell.' Jacqui bruised a leaf with her fingers to illustrate this point. 'Because of this the leaves were once used in less hygienic days for floor covering, hence they are known as one of several "strewing herbs". Reedmace is another member of this fringing reed-swamp community. It grows straight and upright with rather bluish strap-like leaves. Its flower stems can be up to 2·5 metres high, with a cylindrical mass of female flowers at the end – the brown sausage-shaped part – and right above them a smaller mass of male flowers. This is the plant sometimes called "bulrush", the perpetrator of this story being the nineteenth-century artist Sir Lawrence Alma-Tadema who painted this plant instead of the true bulrush or common club-rush when he painted "Moses in the Bulrushes". To this day most people think that reedmace is bulrush. The real bulrush or club-rush, *Scirpus lacustris* is much less dramatic looking.'

We noticed some meadowsweet, *Filipendula ulmaria*, around the edge of the pond and asked Jacqui to tell us about it.

'Meadowsweet is another plant that the Elizabethans used as a strewing herb. The flowers and the foliage have quite different, but pleasant, smells. Meadowsweet was also used to flavour mead, the Anglo-Saxon drink. It is more probable that this is where the name comes from. There is quite a lot of beautiful yellow flag, *Iris pseudacorus*, fringing this pond. In times past this plant was used as a dye plant, black dye being made from the rhizomes. The leaves of yellow flag are a little darker than those of sweet flag. The other tall plant is the great willowherb, *Epilobium hirsutum*, sometimes known by the charming but rather extraordinary name of codlins-and-cream. This is the name of an old-fashioned dish based on baked apples. Some people say that the young shoots, if bruised, smell of apples, but I don't think so. The name may refer to the rosy nature of the flowers, which reflects the colour of codlins-and-cream, the apples in this dish giving the cream a rosy hue. Yet another theory is that codlin is a corruption of the old word "codde", the young capsule which sits beneath the flower.'

Looking out between the fringing plants we could see great rafts of water lilies and amphibious bistort on the water's surface.

'The yellow water lily, *Nuphar lutea*, also known as the wild water lily, is over there on the other side of the pond. The plant is rooted in the bottom of the pond and can grow up to a length of three metres, so it is more likely to be found in deeper water. Its leaves are the largest of any British water plant. The yellow water lily has a very appealing colloquial name – the brandy bottle – as the seed capsules resemble old fashioned spirit bottles and when crushed the flowers smell slightly of alcohol.

'Here almost beside us we have the white water lily, *Nymphaea alba*, which is the introduced domestic variety. It can grow up to two metres in length, but is variable in size. This plant has the largest flower in the British flora and, not to be outdone, has two lovely common names. In Cheshire it is

A coot chick swimming amongst a raft of amphibious bistort, *Polygonum amphibium*, **with their characteristic pink flowering spikes.**

"lady of the lake" and in Wiltshire and Dorset "swan among the flowers".

'There is a lot of amphibious bistort, *Polygonum amphibium*, around the pond. It is sometimes called redleg because of its pink spike-like flowers. You can see a raft of it out in the middle of the pond. That little platform in the centre of the bistort was the second nesting place of a pair of coots which they must have thought safe and sound from predators. However, there was a problem the coots had not counted on. Carp, of which there are a number in this pond, like to lay their eggs in the stems of hornwort, *Ceratophyllum submersum*, where they will be sheltered. But this year there has not been as much hornwort as in previous years. So the carp laid their eggs on the bistort leaves around the coots' nest. It was quite a sight! The carp were in a mass thrashing around the hen, and she took a terrific battering. But she rode out the storm over their great backs as they torpedoed her. Look! There she is with her chicks.'

As we stood quietly the hen coot with her glossy black plumage and distinctive white 'plate' on her forehead came near to us close inshore as she searched for food amongst the thick vegetation. The chicks followed along behind, scurrying across lily pads when they could, trying to keep up with their mother.

We could also see a number of large carp circling in the water. We asked Charlie what they ate. 'The carp feed mainly on the bottom of the pond. They eat water plants, worms and crustaceans.'

We looked around the pond and saw poles sticking out over the water in several places like fishing rods, and Charlie explained what they were.

'Although you can see kingfishers around the nature reserve, they are exceptionally shy birds. So in order that the kingfishers are more obvious to our visitors we put out these poles as perches for them. Nesting birds come here every year. I think they are incubating now because we only ever see one bird at a time. Then, the glorious day arrives when all the young kingfishers emerge! They sit in a line along one of the poles and call to their parents who bring them food.'

We moved on around the pond and Jacqui pointed out the damselflies which were all around us. These exquisite jewel-like creatures, alighting on the reeds and sedges, were out in large numbers on this sunny afternoon. Their lives are inextricably bound up with the ecology of the pond and Jacqui went on to describe the two sub-orders of this insect group.

The empty 'skin' of a dragonfly nymph still apparently clinging to the reed stem which it used to climb out of the water before splitting to moult into an adult. This transformation usually occurs early in the morning on sunny days.

Above **A dragonfly nymph clearly showing its extraordinary modified lower jaw which is hinged and can shoot out to grab its unsuspecting prey.**

Centre top **A female darter or** *Sympetrum* **resting on a pondside log. These dragonflies will often keep returning to the same patch to sun themselves.**

Centre bottom **A common ischnura or blue-tailed damselfly,** *Ischnura elegans,* **with its wings folded over its abdomen in a fashion typical of the damselflies.**

'There are substantial differences between damselflies (Zygoptera) and dragonflies (Anisoptera) although both belong to the order Odonata. A damselfly is smaller than a dragonfly. It is more slender and sits with its two pairs of similar wings folded together vertically over its abdomen. Also, its flight is fluttery and it usually feeds amongst the waterside vegetation. A dragonfly is larger (and takes larger prey), with a stouter body. It has two pairs of dissimilar wings and when it sits it stretches them out like an aeroplane. Its flight is stronger and more direct than that of a damselfly.

'Then, there are two types of dragonfly – hawkers and darters. The hawkers, such as the emperor dragonfly, *Anax imperator*, patrol up and down a stretch of water, snatching their prey – sometimes even smaller dragonflies – as they go. They will defend their territory against all other males of the species and can become quite aggressive. The darters, such as the four-spotted libellula, *Libellula quadrimaculata*, sit on a perch from where they dart out to snatch prey from the air.

'Underwater, in their nymph stage, both the damselflies and dragonflies are much the

A pair of common blue damselflies, *Enallagma cyathigerum*, mating. This order of insects is unique in that males have specialised reproductive organs at the front of their abdomen, to which they transfer their sperm before mating. The males hold the females behind the head whilst copulating as shown in the picture and will often fly together locked in this position.

If you are very patient and are sharp-eyed enough you might catch a glance of the jewel-like kingfisher perched on an overhanging branch, waiting for a fish to come within striking distance.

same, although the damselfly nymph can be distinguished from that of its larger cousins by its more slender shape and by the three leaf-like gills which extend from the back of the body. An interesting thing about their feeding behaviour is that although as adults they tend to use their legs to sweep up food, almost like a basket, in their nymphal stage they use an amazing appendage called a "mask", which flicks out and back. It is rather like having an arm under your chin which shoots out to an enormously long length. Two pincers at the end grasp the prey and bring it back to the mouth to be eaten. When not in use the "mask" folds under the head.

'The best time to see damselflies is mid-summer as most of them are out in June and July. There's the common ischnura *Ischnura elegans* – the damselfly with the blue tip on its tail. This is probably the most common damselfly at Woods Mill, and in most of south and eastern England for that matter. This one may be recently hatched. It seems a bit floppy and could be from the first brood. There's a banded agrion, *Agrion splendens*. That one is a male as it has lovely large apple-green blue bands – one on each wing. The female has greenish-brown wings and a beautiful metallic green body. They tend to flop along in the same way as the meadow brown butterflies.'

We knew that some members of this dazzling group were becoming quite rare and asked Jacqui why this has happened.

'Some species of damselfly and dragonfly are very discriminating as to the kind of water they like. They must have clean, unpolluted water. Those that are becoming rare are doing so because this type of habitat is disappearing. In fact, four or five species have disappeared since the war. Some, however, can tolerate a small amount of polluted water, and the common ischnura is one of these.'

Charlie added, 'What makes these insects particularly vulnerable is the fact that they have such long life histories. The larval stage of the dragonfly may be as long as four or five years. So if a farmer digs out his ditches with a tractor and bucket at a less than five-year interval, the life cycle of these wonderful insects is disrupted so much that they have little chance of survival. Sometimes they come back quickly, if they can find a suitable place not too far from where they were originally. It is much easier for damselflies because their larval stage lasts for only one year. The fringe vegetation in a pond or along a stream may look suitable, but if it has been dredged, it will most likely be barren of these insects.'

Jacqui explained how damselflies and dragonflies lay their eggs.

'There are several different methods. Some simply drop their eggs in the water. Others partially submerge themselves and deposit their eggs underwater on a plant. I remember once seeing a brown aeshna, *A. grandis*, depositing her eggs on a piece of rotten wood – a nice safe home for them.'

Now it was time to take a close look at what was going on underneath the water. We were hoping to see some damselfly nymphs down in the dipping pond. We wound our way down between the verdant grasses and the shade of the willow trees to an elongated pond that was bursting with all manner of life. Charlie told us the history of the 'dipping pond'.

'It was originally the centrepiece for an ornamental garden back in the 1930s. It was neglected over the years, before the Sussex Trust came, and had become entirely silted up from the winter floods. The Trust got their volunteer conservation corps out and rediscovered it. Now we have let it silt up slightly at one end to allow this tremendous variety of plant life to thrive.'

Jacqui began pointing out various pond plants and commenting on their characteristics, including their methods for surviving the winter.

'I think it's quite interesting to consider how water plants do survive the winter. Water lilies, for example, die back each winter and exist until spring in the form of their very thick rhizomes at the bottom of the pond. When spring comes they grow new leaves. This plant, frogbit, *Hydrcharis morsus-ranae*, sends out little winter buds late in the summer. They sink to the bottom of the pond and live in the mud. When the water begins to warm up in the spring, they metabolise their stored-up starch and become lighter in weight. Because they contain oxygen they float to the surface as tiny plantlets and start all over again. This plant has another method of ensuring survival throughout the winter: it produces seeds which live at the bottom until spring when they float to the surface. In Britain they rarely set seeds. The water soldier, *Stratiotes aloides*, has its own amazing technique for getting through the winter. By the end of the summer its leaves become coated with a chalky substance. This soon makes the plant so heavy that it sinks to the bottom. In the spring new leaves, without the chalky coating, are formed. Because they are light, they rise to the surface, to begin the cycle once more.'

We heard the 'kr-r-rk' of a moorhen and Charlie commented, 'This pond generally supports one pair of moorhens and a pair of reed buntings, which like damp areas.'

Jacqui continued with her survey of the water plants.

Have you ever been waved at by a pond-skater? This member of the genus *Gerris* is showing how adept it is at 'skating' on the surface of the pond. It uses its long middle pair of legs to row across the surface and the back pair as a rudder.

Opposite top **An underwater photograph of a back-swimmer,** *Notonecta,* **just under the water surface replenishing its bubble of air before diving again in search of its prey. It uses its long fringed back legs as paddles.**

Opposite below **Although spiders are largely a terrestrial group of animals you might come across one whilst pond-dipping. This will almost certainly be** *Argyroneta aquatica.* **This has the fascinating habit of constructing an underwater 'diving-bell', which it keeps stocked with air brought down from the surface trapped in its body hairs.**

'Here we have fringed water lily, *Nymphoides peltatum*. It is an attractive plant to have in a pond and although it resembles a water lily, it is more closely related to the colourful gentians and bogbeans. These are the arrow-shaped leaves of the arrowhead *Sagittaria sagittifolia*. It is related to the water plantain, *Alisma plantago-aquatica*, another pond plant, although the arrowhead's flowers are twice the size. Another plant that grows in marshy areas and pond edges is the bogbean, *Menyanthes trifoliata*. I think "bean" refers to the shape of its leaves which remind me of those of the familiar garden broad beans. Bogbean was used as a cure for scurvy in the Middle Ages. It has had several other uses over the centuries, including the use of the leaves to flavour beer in the North of England. The flowers are very beautiful – it's too late for us to see them now – but they have five petals fringed with white "hairs".

'This particular plant, hemp agrimony, *Eupatorium cannabinum*, is one of those that butterflies like very much. It is quite tall with reddish stems and tiny pink florets. It adds colour along streams and beside ponds, particularly later in the summer.'

'There are a number of other plants around the pond including creeping buttercup, *Ranunculus repens*, and flowering-rush, *Butomus umbellatus*, which used to be relatively common along waterways, but is now sadly disappearing. I can see some Canadian waterweed, *Elodea canadensis*, down in the water. As you may have guessed, it is an introduction. In 1849 a small female plant was introduced into the Oxford Botanic Garden. It soon escaped and only a few years later this plant was almost completely blocking parts of the Thames. It then went into a period of natural decline, which was lucky because none of the schemes to get rid of it worked. Now it lives much more in harmony with its environment.

'Look, five pairs of damselflies. This pond is absolutely swarming with them. There goes a large red damselfly, *Pyrrhosoma nymphula* – the most common red damselfly.'

Jacqui continued, 'It is worth emphasising that when people go to look at ponds, it is a good idea to sit for a few minutes to absorb the surroundings.' We eagerly sat down in the warm sunlight near the edge of the pond watching the hum of activity around us. Several nets were laid out beside the pond for us to use when pond-dipping. We asked Charlie to outline the equipment needed for

pond-dipping and if we could make some simple equipment ourselves.

'These nets are home-made. It's very easy to do. Take a piece of stout wire, make it into a loop and screw it into a broom handle. The loop is best covered with a piece of strong material. Then, net curtain with a reasonably wide mesh is made into the bag for the net. This is then sewn onto the strong material covering the loop. This will make a very serviceable net and one that will take hard use. If you use nets with different-sized meshes, you will catch different-sized creatures. Then, you need an old ice-cream-carton type dish to put your catch in. It should be wide, not too deep and white. This immediately silhouettes the various animals and helps immensely with identification. You need a good identification book as well. Remember to treat the living things you catch with care. The white container should be about one-third full of water. Keep it in the shade and always return to the pond what you have taken from it, including bits of pond weed.

'Now let's do some pond-dipping. Many creatures prefer to live in one area of the pond: surface film, underwater vegetation, open water or bottom mud. Let's dip in one area at a time and see what we can find.'

We swept our nets across the surface of the pond capturing a number of the water bugs we had seen on the surface. We tipped our finds into the white tubs and asked Jacqui to examine what we had found.

'There are several whirligig beetles, *Gyrinus natator*. These insects have special eyes – half sees above the water and half below. Bugs like this can "skate" on the surface of the water because they have water-proof hairs on their legs. There are still hundreds of whirligig beetles swimming endlessly in circles on the pond, sometimes in clusters. They are probably charging around looking for food, such as insects that have fallen into the water. By the way, the whirligig beetle is a great escape artist from the school aquarium.

'We have also caught a water measurer, *Hydrometa stagnorum*. Notice the extremely thin body. They are quite common and widely distributed. There is a member of the pond skaters, *Gerris*. They are interesting because they don't have larval or nymph stages, the young being born in adult form. Here are some red water mites – tiny, but brightly coloured, and that's really why we notice them. There are several species of red water mite, and not all are found on the surface. We have also caught some lesser water boatmen, *Corixa*. Actually, they usually live in the mud. This one is clamped on to something. I presume it is sucking the juices out of another insect. The lesser water boatman has red eyes and a beautiful blue-green underbelly. It swims the right way up, unlike the greater water boatman or back-swimmer, *Notonecta*, which we also have here. It can cling to the underside of the surface film where it replenishes its oxygen supply. It has large paddle-like legs which really do look like mini-oars. But be careful as this insect can give you a nasty nip. The backswimmer can live both above and below the water. They are good fliers and can quickly colonise a new pond. Well, we have had a good selection from the surface film.

'Now, let's dip in the underwater vegetation, a favourite hiding place for many living things. We should bring up a few bits of the vegetation because many creatures will be attached to it.'

We followed Jacqui's advice, scooping the nets under the vegetation, brushing it to dislodge the animal life.

'We have common newt nymphs, *Triturus vulgaris*, damselfly nymphs and mayfly nymphs. The mayfly nymphs live in the water for a year or more. See the three tails

A delicate damselfly nymph (compare with the much stouter dragonfly nymph on page 34) with its distinctive external gills at the end of the abdomen.

which they all have. If you look closely you can see their vibrating external gills which are arranged along their sides. Like the damselflies, when the mymphs are fully mature they climb out of the water and up onto a suitable plant such as the leaf of an iris or bur-reed. They then moult into the adult form. However, with mayflies it is not quite as straightforward as they have a sub-adult winged form and have to go through another moult before they become the glistening fully adult mayfly. These adults usually only live for a few hours. They quickly mate, lay eggs and die and the cycle then begins again.'

We asked Jacqui why so many insects had their larval stages underwater.

'The pond provides a sheltered environment, the water being a fairly constant temperature, and food is readily available. The larval stage is the soft-bodied stage and obviously moisture is retained when the animal is in water. Insects that come into contact with the air have hard exoskeletons to keep the moisture in their bodies. Ants and bees are two groups that survive on land in all their stages, and their success is mainly because they both have a very complex community life which cares for the developing young.'

Jacqui went on looking at our finds. 'Here are some flatworms or planarians. They really do look like blobs of jelly as they contract and elongate, moving with a gliding motion. Quite often they are found on the bottom crawling over the leaves and stones.

'Obviously, many of these creatures move around from one area of the pond to another. Here are some waterfleas or *Daphnia*. They are not true fleas, but small crustaceans, and despite their small size are easy to see because of their jerky movements. And here are some fish lice, *Argulus*. They are flat and have suckers to hold on to their hosts. If there are enough of them on a fish, they will kill it.

'That's a caddis fly larva (order Trcihoptera), one of the two hundred caddis fly species. They are divided into two groups – the ones that make cases to live in and the ones that don't. This is a case-building larva. Each species makes a distinct type of case – a fascinating subject to study in itself. They use sand and mucus to hold bits of water plant, duckweed, or the short stems of pond weed, together. One type makes its case from

Water lice, *Asellus aquaticus,* **are a common crustacean of ponds and streams. They are mostly scavengers feeding on detritus along the floor of the pond.**

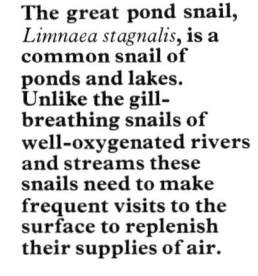

This beautifully marked male great diving beetle, *Dytiscus marginalis*, is an aggressive carnivore and will even try to attack small fish. They are expert swimmers and use their paddle-like back legs to push them through the water. The silvery tip at the end of the abdomen is a bubble of air which is trapped between hairs on the body, this is replenished at intervals by pushing it above the surface of the pond.

The great pond snail, *Limnaea stagnalis*, is a common snail of ponds and lakes. Unlike the gill-breathing snails of well-oxygenated rivers and streams these snails need to make frequent visits to the surface to replenish their supplies of air.

snail shells. You can get them to leave their case by sticking a stem in the back opening of the case – it's open so water can circulate through the case. It doesn't like being tickled so it will crawl out. But it will always find its case again and go back in. If you take the old case away and leave some little coloured glass beads as the only potential building material, it will build a new case from the beads.

'Some of the species that build cases out of plant material often leave a long stem sticking out at the back. This may be defensive, to make it slightly more difficult for a predator to eat it. The main enemies of caddis flies and their larvae are sticklebacks and the larvae of great diving beetles, *Dytiscus*, of which there are a number in this pond. Carp and tench eat the flies in great numbers. A lot of them will have already emerged from their cases by this time of year. Caddis fly larvae usually live in healthy water, but some members of this family can survive in slightly polluted water as well. They have little competition there and can occur in huge numbers.'

Between us, we had collected several types of snail. Jacqui and Charlie looked at them.

'This one is a great ram's horn snail, *Planorbis corneus*. It is fairly large with a black body. We have a bladder snail, *Physa*, and here is a wandering snail, *Limnaea pereger*. It has a huge first whorl with a tiny spire on top. These snails have to visit the surface of the water every so often to refill their so-called "lung", which is just inside the aperture. They are pulmonates, which means that they do not have an operculum to close their shell and that they breathe atmospheric air through their "lung". The other main group are the operculates, which include the freshwater winkles and valve snails. These can close their shells and breathe by means of gills, mostly living in well-oxygenated rivers and streams.

'Leeches are true worms or annelids and are very common in freshwater habitats. They spend a lot of their time attached to underwater plants. Leeches, such as the medicinal leech, *Hirudo medicinalis*, have

A small bottom-feeding pea-cockle, *Pisidium*, clearly showing the extended foot at the front of the shell, which it uses to move around, and the siphon at the back of the shell which is used to draw in water so that particles of food can be extended and also to eject the strained water.

powerful suckers which they use to draw blood from their vertebrate prey. Fish leeches, for example, *Piscicola geometra*, attach themselves to fish and may cause their death.'

After returning these creatures carefully to the pond we tried dipping in the more open water. You have to be rather quick if you want to catch some of the animals of this region. However, with Charlie's expert help we soon had a nice stickleback. We carefully tipped it into the tub.

'This one is a ten-spined stickleback, *Pungitus pungitus*. Its name is derived from the row of ten short spines along its back. Its diet is quite varied and includes insect larvae, worms and crustaceans. The male actually builds a form of nest during the breeding season and uses it to coax several females to lay their eggs in it. He then guards the eggs until they hatch.

'Yesterday we netted a three-spined stickleback, *Gasterosteus aculeatus*, with about twenty fish lice on it. It was in a very

sorry state. In May and June the three-spined males have spectacular red throats and bellies.

'At the other end of the scale in the open water are myriads of microscopic creatures and plants that we cannot see – the plank-tonic animals and phytoplankton or algae. The algae are the beginning of the food chain within a pond. They are consumed by animal plankton and larger creatures such as the water fleas. These in turn are eaten by other animals such as the insect larvae.

'There are some bloodworms, bright red in colour due to the haemoglobin in their bodies. Because of its affinity for oxygen they are able to utilise the little that is available. The name "bloodworm" is used to describe both the larvae of midges and the quite unrelated Tubifex worms, which are true worms. Bloodworms are an important source of food for small predators. You can also find pea-shell cockles, *Pisidium* – here's one – and orb-shell cockles, *Sphaerium*. They filter detritus and organic matter from the water. And here is a red water mite, *Limnochares aquaticus*, one of the red water mite species that lives in the mud. Freshwater shrimps, *Gammarus*, like this one, are scavengers. They feed on decaying plant and animal material on the bottom. You can see quite clearly when it goes underwater that it has a little blob of oxygen on its tail. Some kind of hair mechanism holds it there. We've also caught other animals we have seen elsewhere in the pond, but which really prefer the mud on the bottom, such as flatworms and the lesser water boatman. Although there is little oxygen, there is a plentiful food supply in the form of decaying matter, making the bottom mud appealing to a number of creatures.'

We could have gone on for hours watching life in and around the pond and dipping in the various areas of the pool, but the after-noon was slipping away and it was time to leave. We returned our catches to the water along with the bits of pondweed and walked back to the mill.

Jacqui concluded, 'We have been fortunate to see such a wide spectrum of plant and animal life here at Woods Mill, but they really are things any keen observer could see in or around a healthy pond almost anywhere in Britain. We have seen something of the complex inter-relationships that exist in a pond and have appreciated the beauty of the pond and its creatures. It would be a shame if this type of freshwater habitat were ever allowed to vanish.'

An upland river walk

In the grounds of Kindrogen Field Centre there is a seat with a plaque which records the fact that Queen Victoria took tea at that place overlooking one of the headwaters of the River Tay. There could be no better place to do just that or to take a waterside walk, and there could be no better guide than Brian Brookes, warden of the centre. As he bounds along the bank, his immense enthusiasm and knowledge bubbles over almost as fast as the clear water, which flows over the waterfalls. Think of the problem the aquatic animals have of living in such a place, battered by the rushing waters and in danger of being swept away by spates or even crushed by moving boulders. Yet life in such tumbled waters has its compensations. There is plenty of oxygen, dare I say, always on tap and a good supply of food washed in from the banks or from upstream. You couldn't want a better start than a mountain stream walk to your adventure through waterside Britain.

A classic view of an upland stream cutting its way through the Scottish mountains in a characteristic V-shaped valley.

Information

Stages in the development of a river

Classifying lakes and ponds is a relatively easy task when compared with rivers, as lakes are limited to one place and have distinct boundaries and catchment areas. Rivers, on the other hand can run for many hundreds of miles, passing through radically different landscapes and draining different types of substrate. A simple picture of the development of a river from its source to the sea is to imagine it as containing three distinct zones: a youthful or mountain zone; a mature or foothills zone; and an old age or plain zone. These broad categories, however, fail to reflect the fact that many rivers can change from one apparent type of regime to another and then back again, depending on the immediate topography of the region they are passing through. Because of this, it is general to classify a river according to a series of recognizable zones that reflect the plant and animal life within the river. For instance, a river could be divided up by the species of fish present. This could produce four distinct zones: a headwater zone, where the stream is small and torrential and contains no fish; a trout zone, where trout is the predominant fish, here the water would usually still be torrential and subject to spates, with large boulders and rocks forming the bed of the river; a grayling or minnow zone, where these fish become dominant, here there would be deeper water and, although still fast flowing, silt could accumulate in patches; a coarse fish or bream zone, where the flow is slower and the river is probably meandering with large amounts of silt being deposited. A similar classification can be made using plant types with the sole presence of mosses defining the first zone, the presence of certain types of *Ranunculus*, the second, the inclusion of *Sparganium* and *Potamogeton* species, the third, and finally the last stage would be characterised by the increasing prominence of a range of flowering plants replacing *Ranunculus*.

Broad geographical stages in the development of a river

youthful
or
upland
stage

mature
or foothills
stage

old age
or
plain
stage

Upland river plants and animals to look out for:

Plants

True upland and mountain streams have very little plant growth other than mosses, such as *Fontinalis antipyretica* and *Eurhynchium rusciforme*, as the flow is very strong and often erratic with frequent spates. In slower sections where some silt is deposited a few higher plants can obtain a foothold. These might include:

alternate-flowered water-milfoil,
  Myriophyllum alterniflorum
intermediate startwort, *Callitriche hamulata*
water crowfoot, *Ranunculus aquatilis*
red pondweed, *Potamogeton alpinus*
reed canary-grass,
  Phalaris arundimacea (on margins)
lesser spearwort,
  Ranunculus flammula (on margins)
floating sweet grass, *Glyceria fluitans*
Potamogeton x *sparganifolius* (in Scotland)

Sites

These types of streams are obviously very much a feature of the upland areas of north and west Britain and the rivers that flow through these regions all have their fast-flowing upland tributaries. The tributaries of the **Tay** and the **River Spey** in Scotland both have excellent stretches, although it must be remembered that public access to some of these rivers is limited as they have important salmon fisheries. It is possible to see salmon going through a fish 'pass' on their way to their spawning grounds at **Pitlochry, Tayside.** In Wales there are many beautiful areas with upland streams, such as at **Snowdonia National Park** and the **Brecon Beacons. Taf Fechan** is a wooded river valley managed jointly by the Glamorgan Naturalists' Trust and Merthyr Borough Council. The **Llugwy valley** between Capel Curig and Betws-y-Coed shows many of the features of an upland river. In northern England the upper valley of **Tees** has some superb walks including the waterfalls at High and Low Force.

Animals

flatworm, *Crenobia alpina*
freshwater shrimp, *Gammarus*
river limpet, *Ancylus fluviatilis*
Stoneflies, Plecoptra
  Perla
  Dinocras
  Leuctra
Mayflies, Ephemeroptera
  Ecdyonurus
  Rhithrogena
  Baetis
Caddis-flies, Trichoptera
  Hydropsyche
  Philopotamus
  Agapetus
black fly, *Simulium*

Fish

salmon, *Salmo salar*
trout, *Salmo trutta*
stone loach, *Noemacheilus barbatulus*
minnow, *Phoxinus phoxinus*

Birds

common sandpiper, *Tringa hypoleucos*
dipper, *Cinclus cinclus*
grey wagtail, *Motacilla cinerea*

Mammals

otter, *Lutra lutra*
American mink, *Mustela vison*

An upland river walk
with
Brian Brookes

When you take a walk along the banks of one of Britain's lowland rivers and savour the apparently tranquil scene, it is worth remembering that most of our major rivers have a very different beginning. Probably they start their long journey to the sea from springs or snow meltwaters high in mountains and, for a while, will be characterised as a cascading brook cutting a path down through rocky uplands. These early stages of development have their own distinct associations of plants and animals and have the added bonus of sometimes being found in magnificent scenery. In order to find out more about these upland streams we travelled to the mountains of Scotland to visit Brian Brookes, warden of the Kindrogen Field Centre in Perthshire. In the glorious July sunshine, Brian led us out to the headwaters of one of the tributaries of the River Tay, right up in the heart of the Scottish Grampians. As it turned out, a good road followed the river valley and so, after only a short walk, we found ourselves standing amidst towering mountains scanning miles of heather-covered moors and listening to the distant calling of grouse echoing across the glens. A few yards down a slope from us was a sparkling stream but the real beginnings of the river were right at our feet. We sat down on a great granite boulder whilst Brian began to unfold the story of the headwaters of this particular stream.

'At the moment we are about 2000 feet (625 metres) above sea level and here we have found one of the many places where there is water coming down the hill and out of the ground in the form of springs. Now, one of the first things to look for, when you find a spring like this, is the type of rock that is predominant in the area as this will affect the richness of the water. Interestingly, this area has a very mixed collection of rocks. Some of them are very acidic, like this schist; there is a lot of granite but there are also bands of limestone. This is very important as where the water has been in contact with this limestone it will have dissolved some of the minerals and will consequently be relatively rich, whereas in other places the chemicals in the rock are largely insoluble and the waters at the spring head will lack any mineral nutrients for plant and animal growth. It just so happens that the spring we have before us now is very rich in lime and other chemicals and this shows immediately in the vegetation that is growing around the spring. There are great cushions of mosses actually at the very

A spring at the headwaters of our river. This one has had a rather colourful start! The bright red colouring is a deposit associated with the oxidization of ferrous compounds brought to the surface in the spring water. A very ancient form of bacteria is involved in this process and derives its energy from the conversion. The great clumps of mosses around the spring indicate the mineral richness of the water.

point where the water is seeping out of the ground. There is also a great diversity of higher plants such as sedges, rushes and grasses. Particularly eye-catching is the yellow mountain saxifrage, *Saxifraga aizoides*, which is forming a ribbon down each side of the stream. At this time of year it has these beautiful yellow flowers with orange spots near the base of the petals. It is a very good indicator of calcareous influences. If there is lime in the water we are assuming that we probably have quantities of other important elements for plant life such as magnesium and iron. Indeed, as far as iron is concerned we can actually see places where there are considerable rusty brown deposits associated with the oxidization of ferrous compounds. Bacteria will be involved in this process and will themselves be deriving energy from the conversion of ferrous to ferric iron.

Further into the valley we can see the stream itself which is bubbling and crashing over boulders and pebbles. The bright yellow flowers are yellow mountain saxifrage, *Saxifraga aizoides,* **which gives an indication of the lime-rich nature of this particular catchment area.**

'Looking around we can see some other plants which indicate the very rich nature of this spring. There is a lot of totty or quaking grass, *Briza media*, and a common plant of our lawns which you might not expect to see up here – *Bellis perennis*, the daisy. In a place like this, however, they are an indication of limey conditions. There are also some sedges such as this tawny sedge, *Carex hostiana*, and right at our feet are the tiny white flowers of fairy flax, *Linum catharticum*.'

To one side of the spring the ground was more open and muddy which on closer inspection showed the tracks of many animals. We asked Brian what could have caused such a trampling of the spring.

'Well, it's an absolute quagmire, isn't it? It is in fact another spring area and again the water must be rich in nutrients as the surrounding vegetation is quite luxurient. Considering we are surrounded by extensive areas of pretty tough old heather, this little oasis of lush grasses and other plants is going to attract grazing animals. In this part of the world the biggest and most important animals, apart from the sheep, are the herds of red deer. So what we can see here is ground that deer have churned up while they have been grazing on the succulent vegetation. In the summer the hinds and the stags form separate herds. The hinds, which will have with them the calves that were born in June, form large herds of maybe one or two hundred animals. They tend to keep a little further away from these main roads. If we went over to the other side of the hill we would probably see a herd down in the glen. The stags, on the other hand, keep in smaller groups and wander a little more. So probably it is a small group of stags that has made all this mess.

This spring area has been well trampled by a small herd of red deer which has probably been attracted to the site by the comparatively rich vegetation growing around the spring head.

'Later in the year the stags begin to fight each other and go off on their own to collect as many hinds as they can. Each stag then defends his harem against all comers which, as you can imagine, leads to a certain amount of fighting and noise. Also at this time of year muddy areas such as this are used by the stags as wallows. They deliberately roll over and over on their backs in the wet mud and coat themselves in a dripping layer of ooze. I'm not sure whether this simply cools them off or makes them appear more frightening to aggressors during the fighting.'

We made our way carefully around the spring and headed a short distance down the slope to the main stream. However, before we looked at this in any detail Brian pointed

out that the other side of the valley was worth looking at first. It certainly looked damp and we clambered over to take a closer look. Brian explained why he had brought us across.

'This area of the hillside is a most exciting place for a naturalist. It is what is known as a calcareous flush, which means that it is an area with water rich in minerals seeping out of the ground, like the spring that we have just looked at but much more expansive. An appreciation of the importance of these areas really takes you on to the whole question of the origin of a stream or river. In the first instance, precipitation in the form of rain or snow is obviously the start of the whole process. The water in the stream is derived either directly from this rainwater falling into it or from water that has drained laterally from the surrounding catchment area. The bulk of the water arrives via this second route and therefore the type of soil around the stream is very important. In this area the soil is rich and limey, but just a little further down the soil is sandy and covered with heather and therefore the input into the stream at that point in terms of nutrients is going to be far less. There are many plants that tell you immediately whether you are on acid or limey soil. Further down on the acid side is hard fern, *Blechnum spicant*, bilberry, *Vaccinium myrtillus* and heather, *Calluna vulgaris*, whereas here we have more of the yellow mountain saxifrage, eye-bright, *Euphrasia*, and alpine lady's mantle, *Alchemilla alpina*. Also here is a species of cotton grass with a wide pale green leaf called *Eriophorum latifolium*. This type of cotton grass only grows in calcareous damp areas, whilst further along in the acid area you can see another type of cotton grass, *Eriophorum angustifolium*, which is common in acid and boggy areas and has long narrow and much darker green leaves. So if you can differentiate between these two types of cotton grass you can immediately say whether you are by a base-rich or an acid site.

'If we just look at a couple of the plants here I will be able to show you an interesting reproductive mechanism that a lot of these upland plants use. Here we have some viviparous fescue grass, *Festuca vivipara*, and instead of producing flowers at the top as one might expect, it has these little plantlets, each of which can just break off and get blown or washed away and immediately start to grow into an identical plant. The alpine bistort, *Polygonum viviparum*, has it both ways as it produces conventional flowers at

On the opposite side of the valley is an area known as a soligenous or calcareous flush, which, in contrast to the predominantly acid, heather-covered fells is carpeted in a rich mixture of sedges and other plants. Here we can see the cottony-seed heads of the broad-leaved cotton-grass, *Eriophorum latifolium.*

The flowering spike of the eyebright, *Euphrasia scottica,* **one of the many micro-species of this beautiful little plant.**

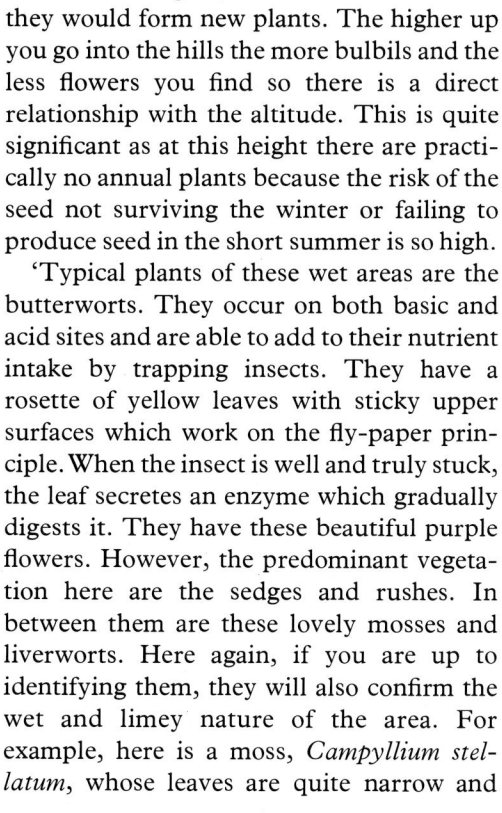

A close-up view of the moss, *Campyllium stellatum,* showing the golden-yellow star-like tips to the fronds.

The alpine bistort, *Polygonum viviparum,* clearly showing the conventional flowers at the top of the flowering stalk and the little bulbils in the lower part. This mixed approach to reproduction gives the plant a better chance of producing offspring in the short unpredictable summer season.

the top of the inflorescence and little bulbils lower down. Again, if these were broken off they would form new plants. The higher up you go into the hills the more bulbils and the less flowers you find so there is a direct relationship with the altitude. This is quite significant as at this height there are practically no annual plants because the risk of the seed not surviving the winter or failing to produce seed in the short summer is so high.

'Typical plants of these wet areas are the butterworts. They occur on both basic and acid sites and are able to add to their nutrient intake by trapping insects. They have a rosette of yellow leaves with sticky upper surfaces which work on the fly-paper principle. When the insect is well and truly stuck, the leaf secretes an enzyme which gradually digests it. They have these beautiful purple flowers. However, the predominant vegetation here are the sedges and rushes. In between them are these lovely mosses and liverworts. Here again, if you are up to identifying them, they will also confirm the wet and limey nature of the area. For example, here is a moss, *Campyllium stellatum,* whose leaves are quite narrow and

nearer to the source of the stream you go the less variation you have as the water coming out from the rocks is of a fairly constant temperature. So although the water feels cold today, in winter it may still be flowing when the rest of the landscape is frozen solid. Some of the animals that are found in these headwaters rely on this constancy of temperature. In particular there is a little flatworm which we might be lucky enough to find here. What we need to do is search among the stones on the stream bed, especially the flat ones sitting on sand and gravel which have a current of water underneath them. What we are looking for is a little flat worm-like animal which will be on the rock surface. Ah, there we have a very, very small one. It is a dirty creamy-ochre colour with sharp ear-like points at the front. It is gliding very slowly across the surface. It is a very primitive little animal and belongs to one of the flatworm families which includes the tape-worms and liverflukes. This is a free-swimming animal called *Crenobia alpina*, and it is very fussy about the temperature of the water in which it lives. It is essentially an animal associated with very cold arctic-type tundra conditions which would have been prevalent here in the immediate post-glacial period about 10000 years ago. In places the

spreading, making these bronze or golden-yellow stars at the tips. And there is a lovely little flat thalloid liverwort with a purple frill around the edge; it is called *Preissia*.

'This is an interesting plant if you are interested in the evolution of plants – the club moss, *Selaginella selaginoides*. It is actually more closely related to the ferns than the mosses. If you move up the evolutionary scale this plant is the first one you come to that produces spores of two kinds – big ones that form the female part of the life-cycle and little ones that are the male part. In lower plants the spores are only of one kind.

'Whilst we are here right at the headwaters it would be a good point to discuss an important physical feature of the stream itself. If you consider that the water has just emerged from these springs, once it is in the stream and out in the open it is going to be subject to rapid changes in temperature. On a hot summer's day like this the rocks will become quite warm and the water tumbling over them will become warm as well. At night the opposite will happen. Further to this, the temperature changes with the seasons can also be very extreme. But the

The common butterwort, *Pinguicula vulgaris*, is a typical plant of the wet areas of north and west Britain. They are able to add to their nutrient intake by trapping insects on the sticky upper surfaces of the leaves. The leaf secretes an enzyme which gradually digests the insect.

conditions are still similar – obviously there is more vegetation now but the water temperature is probably comparable and so this relic of the glacial fauna has managed to survive in the headwaters of these streams.'

We placed the rock carefully back in the stream and moved on down the valley to a nearby bluff. Having looked in detail at the micro-habitats afforded by the headwaters of the stream we now altered our gaze to the overall landscape. To say that it was impressive was certainly an understatement but what were the forces that had formed it and in particular what part did our stream play in the overall pattern?

'First of all it must be remembered that the underlying rocks are extremely ancient – over 350 million years old – and that the mountains have been formed by a series of tremendous upheavals, producing durable metamorphic rocks. But these rocks only form the skeleton of the landscape. In order to understand the scene we see today we have to look at a much more recent time, geologically speaking, when the land was under ice. When you look at the view up here you will notice that there is very little solid rock visible and that most of the landscape is actually covered with a deposit that is hiding this rock. We are looking in fact at a landscape which has been produced by glaciation. If we go back to the depths of the ice-age this area would have been covered by a thickness of ice between a quarter and half a mile in depth. Certainly far higher than the tops of these mountains. So we have to think in terms of a great ice sheet rather than individual glaciers. This sheet slowly

radiated outwards from the main area of precipitation in Scotland which was then, as now, the western side of the Highlands. So material from this western region would gradually have been pushed outwards to be eventually deposited all over these hills. As the climate improved the ice sheet began to thin and at one stage the tops of the hills would have begun to poke out through the ice-sheet and the ice movements be guided by the shapes of the intervening valleys. Gradually the sheet would have broken up into a series of glaciers and it is these glaciers, still propelled by the weight of ice behind them, which went down glens like this and gouged out the wide U-shaped flat-bottomed valleys, cutting off any spurs as they went along.

'Eventually the climate improved to the point where the movement of the glaciers itself began to slow down until it became stagnant and began melting away. At this stage, there was a lot of water running from the glacier and from ice further up the valley which would have brought more material down and deposited it in between the chunks of ice and along the sides of the valley, particularly as a gap opened between the glacier side and the valley. The stones in these deposits are very smooth and rounded, confirming that they were transported by water. If we were to look at material higher up on the hill slopes which was deposited straight out of the ice we would find that it was much more angular and sharp.

'So looking south we can see the wide flat-bottomed valley we have today with the river running in the centre. However, off to our

56

left is a very different shaped valley. This one is much narrower with a V-shaped cross-section. The gradient is quite steep and the stream is flowing quite rapidly down the valley centre, tumbling over little waterfalls and around large boulders. As it goes it will be washing away some of the rocks and gradually deepening the valley. So this valley is the result of erosion by the stream and not by glaciers. Another pointer to this is that the stream is running in a series of zig-zags resulting from deflections from its course produced by obstructions such as hard rocks. We can only see a short distance up the valley as there is an interlocking series of spurs. A glacier would have produced a much straighter valley and removed the spurs.

'Further along where the stream has entered the main glaciated valley it is flowing much more slowly, the gradient is less but is still moving from side to side. The curves are much bigger, however, producing a more sinuous wave-like pattern in contrast to the zig-zag that we had earlier. The bottom of the valley is filled, as these hillsides are, with glacial material and the river is continuing to erode it away. Just below us it has cut into some of the glacial mounds to produce a steep, flat slope which is far more typical of a water-worn slope than one produced by glacial action. However, obviously the over-all shape and size of the valley has nothing to do with the present river.'

Before we moved on, we took one last look at the river meandering down the centre of the glen, the water catching the mid-day sun. What animal life lived amongst the rocks and boulders? Were there likely to be any fish lurking in the deeper pools? These were questions which we were to look at next. Brian took us to a stretch of a nearby river which he has studied in detail over the years and where we could look closely at the river and its life.

'We've now come to a more mature phase of a river although it is still one that is dominated by fairly fast flow and has many boulders of all sizes in its bed. Before we look at the animal life, it would be an idea just to appreciate the range of environments that make up even a short stretch of river such as this. We are standing next to a fairly deep pool with a relatively calm surface whereas just upstream there is quite a rough area with a low waterfall. Also notice how the water flow is much faster in the middle of the river. This has an important effect on the distribution of the boulders and pebbles on the

river bed. For example, you can see that nearly all the large boulders are in the centre and as you move out to the sides they are progressively smaller. This is because in the slower flowing sections some deposition is occurring. However, the flow is still too fast generally for the finer silts to be deposited. Also, remember that a river like this will change substantially from season to season. At the moment it is quite shallow and a great deal of the river bed is exposed but during the winter the water level will be considerably higher, as can be seen from the debris lodged in the bankside tree two metres above the current water level.

'All in all, then, as a habitat for animals this is not a particularly hospitable site when compared with most other freshwater sites. Looking at the water cascading along between the rocks it is obviously going to be difficult to find any animals living on the water surface. Whereas on a pond you might find pond-skaters running around, the river, even now, is running at more than a metre a second, so a pond-skater would have to be moving incredibly fast just to stay where it is. Similarly in the main body of the water only the larger vertebrates such as salmon and trout are going to be able to swim fast enough to take advantage of the environment. So where are the majority of river animals going to be found? The animals will obviously need to anchor themselves to something and therefore, not surprisingly, we have to search the river bed for the small invertebrates. So let's have a search around under these stones.

'Immediately you will notice that there are a lot of algae on the surface of the stones and trapped in between the rocks and pebbles are all sorts of plant debris – pieces of grass, leaves, twigs etc. This tells us that a lot of potential food in fact comes from other systems outside the river. Things fall in all the time; besides the plants, insects drown in the river in quite large numbers. So we can assume that quite a lot of the river animals will be scavengers. However, there are quite complex relationships within the river. For example, we can see a lot of zig-zagging chironomid midges flying above the water. The larvae will be important scavengers on the river bed and in turn will be a major supply of food for the fish population. Perhaps, the best way of beginning to unravel the web of relationships in a river like this would be to look at the life histories of some specific animals.

'Remember when you disturb the river

bed be sure that the boulders are carefully replaced as you are undoubtedly endangering the lives of many small animals if you just haphazardly throw them back.'

Brian had brought along some plastic dishes and a sturdy looking net and soon the water in the dish was writhing with dozens of invertebrates that he had collected. Brian gently picked out a few and told us about them.

'We have two or three dozen insect larvae here, most of which are mayfly nymphs. We can identify them as such because they have these three very long tails. These are sensory organs and in some species like this one they are almost as long as the rest of the insect. The nymph is obviously adapted to living in a fast flowing river as it has a wide flat body and limbs so that it can fit underneath stones and lie close to the rock surface. Its behaviour is also adapted and if you watch them in the

Above **In contrast to the wide U-shaped profile of the glaciated valley this side stream has worn down its own V-shaped valley.**

Opposite **The river at a more mature stage. Although the volume of water has increased the bed of the river is still dominated by large boulders and stones which offer a variety of habitats for river animals.**

Mayfly nymphs of these fast flowing mountain rivers are adapted to cling tightly to the rock surfaces and have developed streamlined profiles in order to lessen friction with the water. Notice the rows of flap-like gills along the abdomen and the characteristic three 'tails'.

overlap and form a sucker with which it attaches itself to the rock surface. So if you were to keep a mixture of *Ecdyonurus* and *Rhithrogena* nymphs in a dish of still water you would find that the latter would soon die as they would be unable to ventilate. There is also some suggestion that the gills might be used in helping the animals to swim and stabilise themselves.

'So if we are to find these kinds of insects in the river we should look in places where the water is moving moderately fast, where there is a good supply of oxygen and also where there is a food supply. This leads us to some of these medium-sized flat stones which have a good current of water underneath. With luck, we will find an insect hanging on upside-down facing the current. Here's a good example, it is a stonefly larvae. It looks at first glance not too dissimilar to the mayfly larvae but does, in fact, belong to a separate order of insects – the Plecoptera. They have two, instead of three, long sensory tails and some of them are quite big. Certain species, such as *Perla bipunctata*, can be as long as three or four centimetres including their tail, and take as long as three years to mature. What we have here is one called *Dinocras cephalotes*. Interestingly, the gills of stoneflies are not on their abdomen, as in mayflies, but are on the thorax at the top of their legs, looking like little pieces of cotton wool. If we put one into still water it can't beat its little fluffy gills so it will start to do 'press-ups'. The shorter the supply of oxygen, the more furiously it will do its 'press-ups' in the same way that a mayfly will beat its gills faster.

'Like the mayflies, the stonefly adults emerge in early summer. The adults are rather drab-coloured but are big and spectacular looking and still retain the two tails. They lay their black eggs in frothy bubbling masses in between stones up above the water level – hence the name, stonefly.

'Our dish also contains some freshwater shrimps, *Gammarus*, which can be found in practically any river or stream in Britain providing the water has a fairly high calcium content. They are mostly scavenging animals although they do eat algae. Their behaviour is interesting in that they respond very quickly to light, for example, in a dish like this they will always swim to the part that is in shadow. This indicates that they are essentially nocturnal, and spend the day hidden under stones. They have a daily dilemma in that they are under the stone during the day but the algae on which they

river you will notice that they do not swim but will run across the stones, always hanging on tightly. In these fast flowing rivers which are constantly tumbling and crashing over the rocks, the water contains a lot of oxygen and the animals that live here tend to be dependent on these high oxygen levels. That is one of the reasons why they don't look too happy in the dish. One of the things they are doing, if you look very closely, is beating their gills, which are in pairs along the side of the abdomen. These 'gills' are ventilating organs and create a current of water over the body so that the insect can absorb as much oxygen as possible. If we were to watch them in the river we would notice that they hardly move their gills as there is sufficient oxygen. This species of mayfly, which is typical of stony rivers, is called *Ecdyonurus venosus*. Together with it we have another closely related mayfly called *Rhithrogena* whose gills have become enlarged and are no longer mobile. They

marus swimming upstream as, given a current, this shrimp will swim into it. This enables them, when the river is not in spate, to make up for the effects of drift.

'We can also see some representatives of another important group of freshwater insects – the caddis flies. The larvae of these flies use material from the river to construct cases within which they live. The case is attached to a silk envelope which they spin around themselves before starting to build. This one has a beautifully constructed case made up of small grains of sand, little pieces of quartz and flakes of mica. Each species has a different type of case and this one is round in section with a slight curve along its length. We can tell which is the oldest end because the diameter of the case increases as the developing insect adds material at the front end. The larva crawls around with its abdomen inside the case and is able to stick its head, thorax and legs out in the front. In these fast flowing waters perhaps a better strategy might be to fix your case to a large rock and wait for the food to come to you. Indeed, in the river here are a lot of *Agapetus* species which do just that. There is another caddis fly, called *Hydropsyche*, which simply spins itself a case of silk. This becomes elaborated into a funnel-like web which catches food carried along in the current, just like an underwater spiders' web. If you have a clear stream you can sometimes look down into the water and see what are apparently crescent shapes on the bed of the river – these are the entrances to their silk nets. But for filter-feeding *par excellence* you want to look into the waterfall areas where the water is flowing fastest. This is where we find the larvae of the black flies, *Simulium*. Quite contrary to many of the other animals they seek out the lightest places rather than the darkest ones and this concentrates them on the outer surfaces of the rock where the water is fastest. They also respond to the current itself, moving towards the fastest flowing water. Here, from glands on the head, they spin themselves a little pad of silk which is attached to the rock. They hook their posterior end, which has great cones with curved teeth, on to the pad and then go limp in the water, extending their mouth parts which are like sieves. There they lie in the current catching food by filtering the water. When they move from one place to another they simply spin another pad and loop themselves onto it and by using the silk pads as a series of stepping stones they slowly get around.

The stonefly nymphs only have two 'tails' and if you look carefully you can see the feathery gills on the thorax at the top of the legs. This is a *Dinocras* nymph, which is typically found in clean mountain streams.

feed at night grows on the top surface. They therefore have to make this rather tricky manoeuvre from the bottom to the top and back again. If you fixed a net in a stream facing upstream and regularly monitored it every hour you would find that very few *Gammarus* were caught for most of the time, but there would be spectacular numbers being swept along by the river at sunset and sunrise when they were making this transition.

'This brings up the subject of drift. No matter how careful you are, at some stage, if you are a small invertebrate, you are going to be swept downstream. For the insects, this is not a very great problem as the adult insect, be it a mayfly or a mosquito, will always fly upstream to mate and lay its eggs, thus compensating for any downstream drift of its larvae. But what do the shrimps do? If you could fix your nets facing downstream you would find that you would still catch *Gam-*

The salmon has an extraordinary life cycle that begins as a fry high up in the headwaters of upland rivers. The young salmon or parr then remains in the river for a year or two before moving to the lower reaches as a smolt prior to going to sea. It will then stay at sea feeding in the rich waters off Greenland for some years attaining its full adult status, before eventually returning to its natal-headwaters to breed.

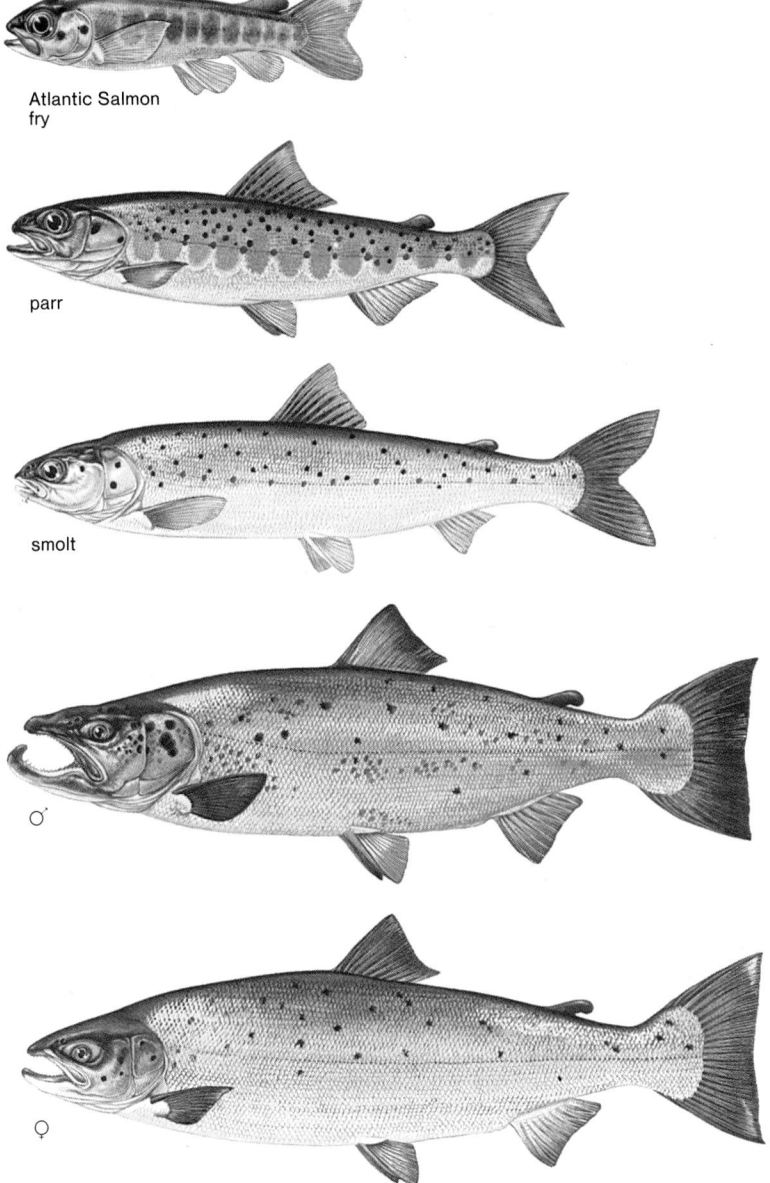

Atlantic Salmon fry

parr

smolt

♂

♀

'These are just some of the many invertebrates that are to be found on the river bed. As we have seen, some, like the *Simulium*, are filter feeders, others, like the big stonefly nymphs are carnivorous and will eat the herbivores and scavengers. But they all in turn will be prey to the larger vertebrates such as the fish, which are near the head of the food chain. A river like this will have large numbers of brown trout, and at certain times of year salmon. The trout, which are entirely freshwater fish, will be feeding and growing in this part of the river but will move up to the headwaters to breed and spawn. These trout will be feeding mostly on the stonefly and mayfly larvae and the *Gammarus*.

'The salmon, of course, are a very different story. However, there will be, even with the river as low as this, a large number of salmon here. But if you were to catch any you would find they were only four to six inches long (15 centimetres). This would be because they would be the very young stages of the salmon, hatched from eggs laid up in the very headwaters of the stream in the middle of winter. They will have moved down to this part of the river as very small salmon called parr and will spend a year or two in the river, feeding and growing very slowly. At this stage they look very much like trout and can be difficult to distinguish from them. They do have, however, very pronounced dark marks all the way down the side which the trout usually do not, but this is not infallible. Eventually they become much less colourful and rather more silvery, with few strong patterns. They are turning from parr into smolt and this is when they move further down the river before finally going to sea. They stay at sea for some years, apparently the time varies from salmon to salmon, feeding mostly around the waters of west Greenland. But in the end they always return to breed. Amazingly they are able to retrace their way right back to the headwaters where they were spawned. The adult salmon will have grown phenomenally in the period at sea – these are the fish we generally think of as the salmon. The cock and hen fish, as they are called, will move up the river always swimming against the strongest currents. When the rivers are low, such as now, they will hang back until there has been some heavy rainfall and then when there is a surge of water down the river, the salmon will begin to move up. At this stage in their life cycle they don't feed as they undergo some very fundamental changes on their return to freshwater. One of these is that the whole of the gut tends to atrophy. This presents a problem for the angler, who has to catch the salmon by annoying it rather more than by tempting it to eat. Sometimes this long journey up the river means that the fish arrive in the spawning grounds in pretty poor condition and afterwards they normally die. These spent fish are called kelts, and in the early part of the spring you can find four or five of these emaciated dark red kelts washed up along a hundred yard stretch along here. Some of them manage to get back to the sea where they regenerate their digestive system and will return to breed again.

'Another group of vertebrates at the top of the food chain are the birds. If you are visiting a stretch of upland river it is always worth looking out for the delightful dipper. They like these rivers, with plenty of boulders to perch on and fast clear water with plenty of invertebrates to feed on. You will often see this dumpy dark brown and white bird flying low and straight over the water and landing on a protruding rock in the centre of the river. There it will stand dipping and bobbing up and down, doing "knee-bends". If you get a good view another point to look out for is its eyes. For the dipper, like most birds, has a third eyelid, a nictating membrane. This shows up white in the dipper contrasting with its dark head. Suddenly it will just pop off the boulder to feed in the

river. However, unlike a wader, which would paddle around poking into the shallower areas, the dipper will actually go under the water and walk on the bed, held down by the force of the water pressing down on its back. Here it moves along upstream turning over stones with its beak and fishing out the crustaceans and insect larvae which are hidden underneath. Dippers often nest in bridges and stonework or occasionally in overhanging trees. They will then construct a very big nest out of twigs. Each pair has a definite stretch of river which is their territory and they patrol this regularly. The number of territories on a river will directly reflect the amount of food available and therefore the productivity of the river.

'Another bird that is typical of these stony rivers is the grey wagtail. If you just get a quick glimpse of one travelling upstream with its typical bounding flight the most noticeable feature, despite its name, is the bright sulphur yellow belly and perhaps the black bib of the male. They hawk over the surface of the river, picking off the adult mosquitoes, midges and mayflies. Like the dipper they breed along the banks of this river, perhaps in a clump of grass hanging down over a boulder or beside a waterfall in a rock crevice.

'Whereas these birds, to some extent, feed, like the fish, on the profuse invertebrate life in the river, they are predators that, in a sense, come further up the food chain and can feed on the fish themselves. It could also be said that they are from partly outside the river ecosystem. If we look along the bank of the river in sandy places or where there's a little muddy backwater we might be lucky enough to see the webbed tracks of an otter. If you are very lucky you might see a stream of bubbles in mid-river followed by sight of a flat wedge-shaped head and long whiskers.

The dipper is a familiar bird of rocky upland rivers and streams. It has developed the ability to actually walk under the water, where it catches insect larvae and crustaceans on the river bed.

But unfortunately there are very few otters and more often than not when you see a large carnivore along a river it turns out to be something rather smaller and darker, much more weasel-like – a mink. These are usually escapes from fur farms and can be a fairly serious threat to river fisheries because they seem to kill a lot of fish and eat very few of them. In contrast the number of otters is so small in this part of the world that the amount of fish they kill is relatively insignificant.'

While Brian was telling us about the otter we scanned the river with the vague hope that we might actually see one. But, of course, we had no such luck. We returned our stonefly nymphs carefully to the river. Brian had certainly shown us some of the rich natural life that can be found in these upland streams and rivers. But an important part of the overall equation was still to be brought in – the effect of that arch transformer of all things – man. We drove downstream to Blairgowrie to discuss some of his uses and abuses of our rivers. Brian took up the story.

'We have now come much father down the river system where the river is much bigger. The volume of water being discharged at this point is something like four or five times the volume that we were seeing further up. The increased volume means that the potential of the river in terms of energy and therefore as a source of power has increased tremendously. Whereas in the glens man used the river to drive small mills for grinding flour, we are now in a situation where a considerable source of energy could be trapped for industrial and semi-industrial purposes. This source of power is the key to the growth of Blairgowrie in that the prosperity of the town was built on the success of the mills along the riverside here. Although they are sadly no longer in use they started originally as flax mills for linen and soon after that were associated with the growth of the jute industry in Dundee. The mill we can see upstream has still got its great water wheel which right until the present day contributed to the power requirements of that mill. For most of the mills, the water power over the years was initially supplemented and eventually largely replaced by coal. However, it is interesting that one mill actually went back in its last days to the river again, using it as a source of power supply to drive its own small hydro-electric turbine plant. So this is a very significant relationship between the river and man.

'Not all of man's relationships are quite so happy, I'm afraid. We do get examples all along the river of its being used for the disposal of waste, rubbish and effluent both domestic and from farms and industry. This has a considerable effect on the river and its life. You can actually gauge to what degree a river is polluted by the types of plant and animals that are living in it. For example, we were only able to find the stonefly and mayfly larvae earlier on because there was practically no pollution at that point. If the pollution was uncontrolled the river could easily be reduced to a lifeless habitat. There are, of course, many types of pollution. Moderate organic pollution from sewage can be regarded in a river like this as a slight enrichment of the river system and many of the animals in the river may well depend upon such organic input for food. We often see a lot of fungi in river water showing as a grey felty growth on the surface of the rocks. This will spread dramatically if you increase the degree of organic matter in the system and it can begin to cover the algal growth in the river itself. Other forms of pollution can

Further down the river with its increased volume, man has utilised the tremendous power represented by the flow of the water. This riverside mill was once powered by a great water wheel.

be much more insidious. One can think of instances of toxic effluent going into a river and having a dramatic effect on the river life. Other forms of pollution can be important in some areas. Even a power station may be using water from a river as a coolant and then discharging warm water into the river. This thermal pollution may have its good and bad points. Mining and quarrying may appear to be innocuous yet the very fine suspended material discharged into a river can, for example, clog up the gills of the fish. A well-aerated river such as this with its rapids and tumbling waterfalls can take a degree of organic pollution because of the high levels of oxygen for biological activity, but slower flowing reaches can become stagnant and the river life severely affected.

'One final point that we should remember is that just as at the headwaters, the chemistry of the water here is dependent on the surrounding land use. The chemicals will depend partly on the geology over which the river has come and partly the soils through which the rainwater has percolated and in turn the use to which man has put the landscape. If it is an arable landscape the farmers will probably have used a great deal of artificial fertilizers and many of these, especially the nitrogenous ones, are likely to be washed away from the fields and will eventually end up in the main river. Application of fertilizer to forestry land will also result in chemicals, such as calcium phosphate, being washed into the river system. All this will immediately affect the nature of the plant and animal systems in the river. So it is important to be aware of all these diverse influences when looking at a river system.'

Brian certainly was not exaggerating about the tumbling waterfalls along the river as we were standing beside a spectacular series of gullies and rapids carved out of sedimentary conglomerate rock. It had been a superb day and despite the fact that we had seen so much, from the mountain-side springs to this magnificent stretch of river lined with mills, we had only really scratched the surface. And I for one am looking forward to visiting these upland rivers again for another look, and who knows, I might even catch a glimpse of the elusive otter.

A lowland river walk

All rivers pass through a process of maturation as they make their own individualistic way down from source to sink in the sea. In their youth, they rush headlong down tearing away at any rocks which dare to impede their progress. In maturity, they meander through flatter country, the combined might of the waters, derived from their tributaries, creating almost as much as they destroy. The pulse of a river reflects that of the landscape while the vegetation along its margins and the wildlife and fish which visit and live within its waters, reflects the health of that same landscape both in natural and man-made terms. Thanks to the growing body of knowledge concerning the life of our rivers, at last many are being made cleaner and gross uncontrolled pollution is a thing of the past – although eutrophication is still an increasing problem. Nigel Holmes is one of our greatest river walkers, recording the facts of change in terms of plant and animal communities. He usually records his data on floppy discs for computer retrival, here he records it all so everyone can understand!

An idyllic lowland river scene with alders shading the far bank and a carpet of water lilies and fringing bur-reeds in the foreground – all classic signs of a river flowing over clay.

Information

Although, as we have seen, rivers can be classified into broad geographical zones, within these there is a tremendous variety of types which are determined by the type of rocks the river is flowing over and localised topographical features. This chapter describes a walk along a lowland river in the Midlands which flows over a gravelly sandstone substrate but also has stretches where it is dominated by a more typical clay substrate. Each of these types has its own communities of plants and animals. These have been studied by scientists to produce a more detailed type of classification where considerations such as substrate, flow, depth and volume of water are all taken into account.

The following diagrams show an idealised cross-section of plants in two types of lowland rivers: those on clay and those on sandstone. The next chapter looks at the wildlife of our richest lowland river-type – the chalk stream.

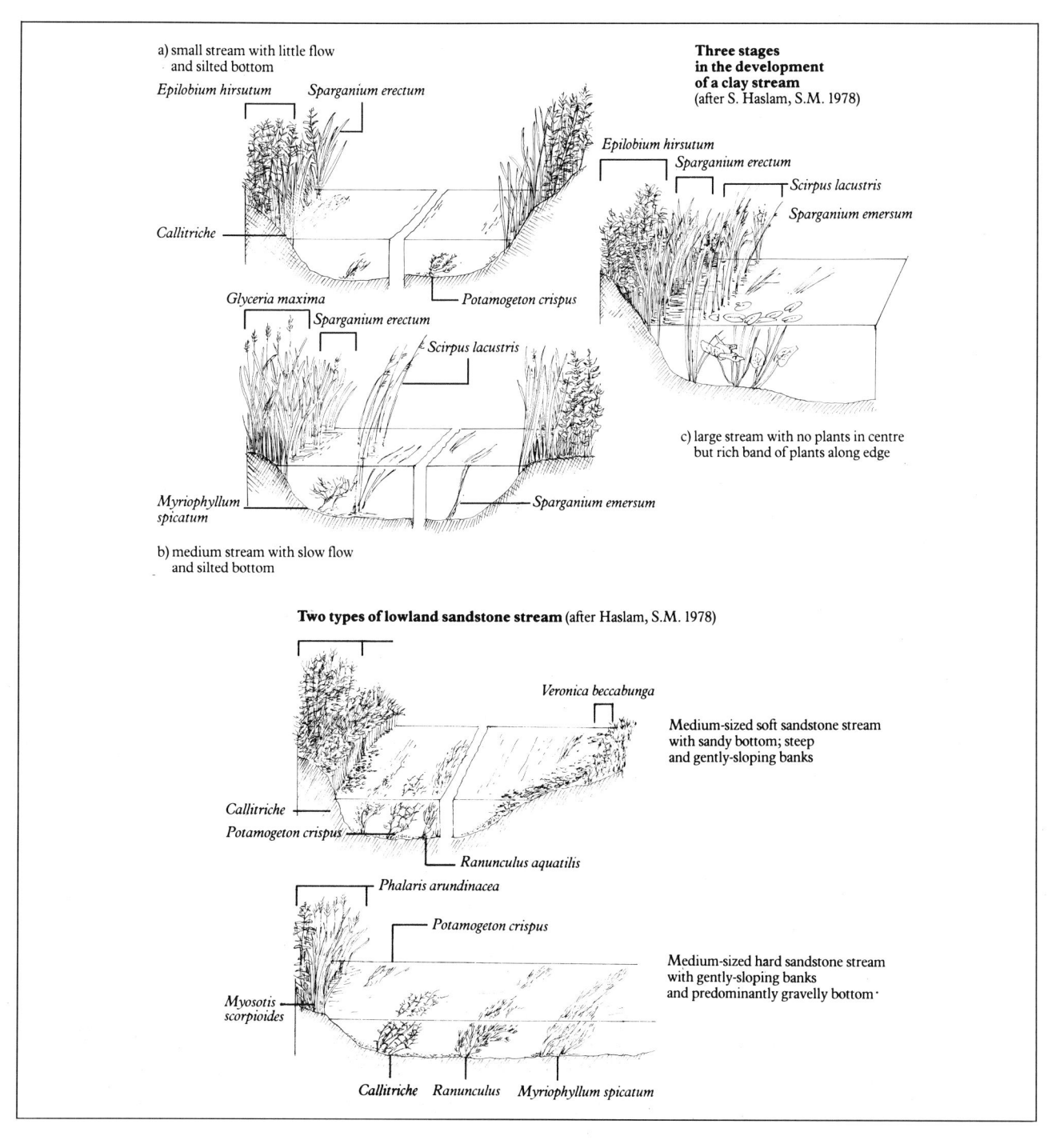

a) small stream with little flow and silted bottom

Epilobium hirsutum   *Sparganium erectum*

Callitriche

Glyceria maxima

Sparganium erectum

Scirpus lacustris

Myriophyllum spicatum

Sparganium emersum

b) medium stream with slow flow and silted bottom

Potamogeton crispus

Three stages in the development of a clay stream
(after S. Haslam, S.M. 1978)

Epilobium hirsutum

Sparganium erectum

Scirpus lacustris

Sparganium emersum

c) large stream with no plants in centre but rich band of plants along edge

Two types of lowland sandstone stream (after Haslam, S.M. 1978)

Veronica beccabunga

Callitriche
Potamogeton crispus

Ranunculus aquatilis

Medium-sized soft sandstone stream with sandy bottom; steep and gently-sloping banks

Phalaris arundinacea

Potamogeton crispus

Myosotis scorpioides

Callitriche   *Ranunculus*   *Myriophyllum spicatum*

Medium-sized hard sandstone stream with gently-sloping banks and predominantly gravelly bottom·

Plants

river crowfoot, *Ranunculus*
yellow water-lily, *Nuphar lutea*
fool's water-cress, *Apium nodiflorum*
water forget-me-not, *Myosotis scorpioides*
great willow-herb, *Epilobium hirsutum*
water-plantain, *Ailisma plantago-aquatica*
branched bur-reed, *Sparganium erectum*
un-branched bur-reed, *Sparganium emersum*
arrowhead, *Sagittaria sagittifolia*
duckweeds, *Lemna*
curled pondweed, *Potamogeton crispus*
perfoliate pondweed, *Potamogeton perfoliatus*
fennel pondweed, *Potamogeton pectinatus*
Canadian waterweed, *Elodea canadensis*
bulrush or clubrush, *Scirpus lacustris*
reedmace or bulrush, *Typha*

Insects

The variety of insects that can be seen during a walk along a lowland river depends on the nature of the river itself and on the vegetation both along the bank and in the water. Some of the most eye-catching insects that you are most likely to see will include:

mayflies Ephemeroptera
    Ephemera danica
    Baetis rhodani
dragonflies and damselflies Odonata
golden-ringed dragonfly, *Cordulegaster boltonii*
banded agrion, *Agrion splendens*
large red damselfly, *Pyrrhosoma nymphula*
common blue damselfly, *Enallagma*
    cyathigerum
alder flies Neuroptera
    Sialis lutaria
water bugs Hemiptera
river cricket, *Velia caprai*
river pondskater, *Aquarius najus*
caddis flies Trichoptera
grammon, *Brachycentrus subnubilus*

Fish

grayling, *Thymallus thymallus*
perch, *Perca fluviatilis*
eel, *Anguilla anguilla*
roach, *Rutilus rutilus*
rudd, *Scardinius erythropthalmus*
dace, *Leuciscus leuciscus*
chub, *Leuciscus cephalus*
minnow, *Phoxinus phoxinus*
tench, *Tinca tinca*
gudgeon, *Gobio gobio*
bleak, *Alburnus alburnus*
barbel, *Barbus barbus*
bream, *Abramis brama*
carp, *Cyprinus carpio*
pike, *Esox lucius*

Birds

great-crested grebe, *Podiceps cristatus*
little grebe, *Tachybaptus ruficollis*
heron, *Ardea cinerea*
Canada goose, *Branta canadensis*
mute swan, *Cygnus olor*
mallard, *Anas platyrhynchos*
tufted duck, *Aythya fuligula*
water rail, *Rallus aquaticus*
moorhen, *Gallinula chloropus*
coot, *Fulica atra*
grey wagtail, *Motacilla cinerea*
reed warbler, *Acrocephelas scirpaceus*
sedge warbler, *Acrocephelas schoenobaenus*
reed bunting, *Emberiza schoeniclus*
kingfisher, *Alcedo atthis*
sand martin, *Riparia riparia*

Mammals

Daubenton's bat, *Myotis daubentonii*
water shrew, *Neomys fodiens*
water vole, *Arvicola terrestris*
otter, *Lutra lutra*
American mink, *Mustela vison*
coypu, *Myocastor coypus* (Norfolk Broads only)

Sites

Britain has many superb lowland rivers, many of which have sections which are of interest to the naturalist and walker. The most famous river of them all is the great **River Thames** which flows gently across southern Britain. Each area of Britain has its own major river systems – for example, the **Severn** on the west, the **Ouse** in the east and the **Tyne** and **Tees** in the north. However, some of the smaller lowland rivers which may not have has as much industrial development along their lower reaches are often well-worth exploring.

The **River Wye** which cuts a sinuous route down along the Welsh Border counties to the Severn estuary has some marvellous stretches. Particularly good areas are to be found around the **Forest of Dean** where the river has cut great gorges into the landscape. **Symonds Yat** is a popular tourist viewpoint.

Some of England's rural scene can be enjoyed at its best at such places as the **Hambledon Valley** in the Chilterns, **Windrush Valley** in Oxfordshire and **Dedham Vale** beside the **River Stour** in Suffolk, countryside immortalised by John Constable.

A lowland river walk
with
Nigel Holmes

The majority of rivers in lowland Britain have been substantially affected by the activities of man who uses them as a source of both water and power supply; as a means of disposing of large amounts of waste products and, not least of all, for navigation from one region to another. Nature, in its way, has adjusted to these many demands and it could be said that man may have created as many new habitats for river plants and animals as he has destroyed. In some areas, where there has been uncontrolled releasing of industrial and sewage effluents into the river systems, this balance has been dramatically upset and the river life has largely disappeared. However, we were fortunate to discover a stretch of river, near to the great industrial conurbation of Birmingham, that has largely escaped these more excessive problems and displays a marvellous diversity of lowland river plants and animals.

It was a thundery day in early August when we travelled up to meet our expert for the day, Dr Nigel Holmes, a member of the Nature Conservancy Council's scientific team, and whose job it is to ensure that the precious heritage of wildlife in Britain's rivers is adequately considered from a nature conservation point of view. Luckily, the storms had largely passed over and we were able to kit up in our wellington boots and macs hopeful of a fascinating afternoon's stroll.

We started our walk from a bridge where the river was no more than five metres (15 feet) wide. On one side of us were marshy pastures where a small herd of cattle were quietly grazing. The river was flowing quite swiftly but, unlike the Scottish mountain stream we had seen earlier in the year, it was fringed not with boulders but with luxuriant stands of reeds. While we surveyed the tranquil scene, we asked Nigel to tell us about the background to this little river.

'This is the River Blythe which rises on the mudstones south of Solihull and comes down east of the Birmingham Exhibition Centre to flow into a larger river, some seven kilometres (four miles) from here. It is a gem of a lowland river that has a reasonable quality of water and which, ironically, flows into what is reputed to be one of the worst polluted streams in the country – the River Tame – which runs right through Birmingham then into the Trent. On a clear day you can look from this gentle rural scene downstream to the power stations and factory chimneys along the Tame.

'You are probably asking yourselves why this stretch has remained unpolluted? This river has perhaps escaped some of the fate of the Tame and the Trent simply because it is too shallow to be navigable by large boats for most of its course and this has probably contributed to the lack of any large industrial sites along its length. There is some treated sewage input from nearby Solihull, but not enough to radically alter its chemistry.

'The type of rock that a river flows over is critical in determining its regime and the Blythe has an interesting mix of mudstone and sandstone overlain in areas with boulder clay. If you are in an area dominated by clay, which is a very impervious substrate, the rain runs straight off the land and into the river

system. Therefore, the level of the river rises and falls exceedingly fast. Some on the clay areas of the Weald, for example, will rise and fall as much as five metres (15 feet). This feature is very noticeable as the bridges are generally very high above the river. Rivers which flow over a porous substrate, such as chalk, are very different as rainwater collects in underground "reservoirs" to be gradually released into the system. If you look at a bridge over a pure chalk stream you will probably find that it is no more than 30 centimetres (one foot) above the surface of the water, reflecting the very small amount of rise and fall. This river has an intermediate flow regime, as the mudstones and sandstones have a certain porosity. However, the

areas overlain with clay are much more susceptible to sudden changes, so the banks are not as shallow as on a chalk stream.'

The plant life was obviously an important component of the natural community and we asked Nigel to tell us what were the most typical plants of our lowland rivers.

'It is very difficult to say what is typical, as, depending on exactly where you are within a river, you will find different assemblages of species. Every section of river provides its own habitat. For instance, as we walk downstream from here we will go from gravel stretches into silted clay areas, where we will find very different groups of plants. However, generally the plants that are typical of lowland rivers are the ones that will grow in

The River Blythe near Birmingham showing that you don't have to go to the Constable country to find some beautiful stretches of lowland river. The dark green stripes in the water are beds of *Ranunculus*.

The familiar leaves and fruits of branched bur-reed, *Sparganium erectum*, **one of the most typical fringing plants of our lowland rivers.**

tight-knit rigid rosettes of leaves. So we can see that not only flow but also the nutrient status of the water is important in determining which plants you can find.

'In very slow-flowing rivers and backwaters, you can find plants that are not rooted at all, such as the duckweeds, *Lemna*, which simply float on the surface and are able to reproduce fast enough to replace numbers lost downstream. These may also be found in faster flowing lowland rivers at the edges where a marginal fringe of vegetation protects them from being washed away. In southern England, a floating fern, *Azolla filiculoides*, may often be seen. This has fine hairs covering its surface which hold raindrops that glisten in the sun.

'An important type of plant that one finds in lowland rivers are those which, although rooted underwater, have a substantial part of their growth above the water surface. Of the smaller emergent plants, which will have some leaves underwater and some above, an interesting example is the river water-dropwort, *Oenanthe fluviatilis*. I mention this plant as it is an important species in the British Isles, which requires conservation consideration. It is a common plant in southern lowland rivers only, especially in those which have a base-rich element, yet it is probably the only aquatic plant in Great Britain which is more common here than anywhere else in the world. This dropwort is confined to Europe and contrasts with many of our other plants deemed to be "rare" or "endangered". They are often other plants which may be just hanging on here, yet may be common or even pest species elsewhere in Europe, or in other continents.

'The fringing reeds are another example of the emergent river plants. If the river flows through what used to be a wetland area its banks may now constitute the only remaining wetland environment. We often find on these banks what might be called relic species of the former general wetland plant community. Plants like the common reed, *Phragmites communis* and reedmace or bulrush, *Typha*, are examples of this. There are many more species that we could discuss and we shall be looking at some as we make our way downstream.

'Looking down the bank of the river, an obvious feature are the willow trees, *Salix*, that are present all along the edge. The two most common willows on the edges are the white willow, *Salix alba*, and the crack willow, *Salix fragilis*. Both are frequently

deep water – some of which never reach the surface. The pondweeds, *Potamogeton*, are an example of a family that are rooted in the river bottom and can either remain totally underwater or have leaves that reach the surface. The broad-leaved pondweed, *Potamogeton natans*, has both floating and submerged leaves. Another group of plants that grow under the water are the starworts, *Callitriche*. This group has different species growing according to the nature of the water. If you have a very acid lowland stream, such as you might find in the New Forest of Hampshire, you will find *C. hamulata*, which has long linear leaves with spanner-like tips. In the more sluggish sections of a river you might find *C. stagnalis*, which has more oval-shaped leaves. Where the water is more basic or calcareous and still reasonably sluggish, you find *C. obtusangula*, which has very planted. These are very much part of the lowland river scene and the ones here have been traditionally cut, as they are mostly pollarded. This means that the branches have been cut back regularly at about two to three metres (six to nine feet) above the ground. This provided a supply of poles for rural industries and kept the shoots out of reach of the browsing cattle. In recent years, with the decline in these industries and the shift in farming emphasis, the cutting has usually been carried out by the water authority as the overhanging branches might create a hazard to obstruct the flow of the river. Also, this remedial cutting might prevent the whole tree from falling in. The other main riverside tree is the alder, *Alnus glutinosa*. This apparently sombre tree has distinctive pendulous male and round female catkins which appear before the leaves. The seeds in the woody cone-like fruits are sought out by flocks of siskins and redpolls during the winter months.

'Different trees have different values for wildlife. For instance, the otter is much more likely to have a holt in an overhanging ash or sycamore, because they tend to form large boles which bow out so that the otter can

Where there are bays and obstructions along the river course some of the plants of slower flowing waters are able to colonise the water surface, such as the minute duckweeds, *Lemna.* **Further along the river are some pollarded willows which are another common feature of these rivers.**

Seen in close-up the metallic greens of this female banded agrion, *Agrion splendens*, are quite breath-taking. This lovely insect can be seen fluttering along riverside vegetation for most of the summer.

excavate a small cavern underneath in which to hide. Insects and fish, on the other hand, can use the shelter of the more matted root systems of the willow and alder.

'Trees also often provide the only permanent, solid objects at the water level in lowland, slow flowing rivers. They therefore provide the base on which aquatic mosses attach themselves. The two most common ones are *Fontinalis antipyretica* and *Rhynchostegium viparioides*. These plants will be typical on man-made structures such as weirs and bridges.'

Whilst Nigel had been describing some of the plant life, a brightly coloured damselfly

had been flying in and out of the fringing reeds, its bright metallic colours catching the afternoon sun. We asked Nigel to tell us some more about the insects of the rivers.

'The most obvious insects and probably the most interesting to the walker and amateur naturalist are the dragonflies and damselflies, Odonata. Some of the larger dragonflies have territories which they patrol regularly, fighting off any intruders. As they feed in flight during daylight, they provide spectacular displays. As they are so noticeable they are very good indicators as to the health of the overall habitat. In recent years we have lost four of the forty or so species in

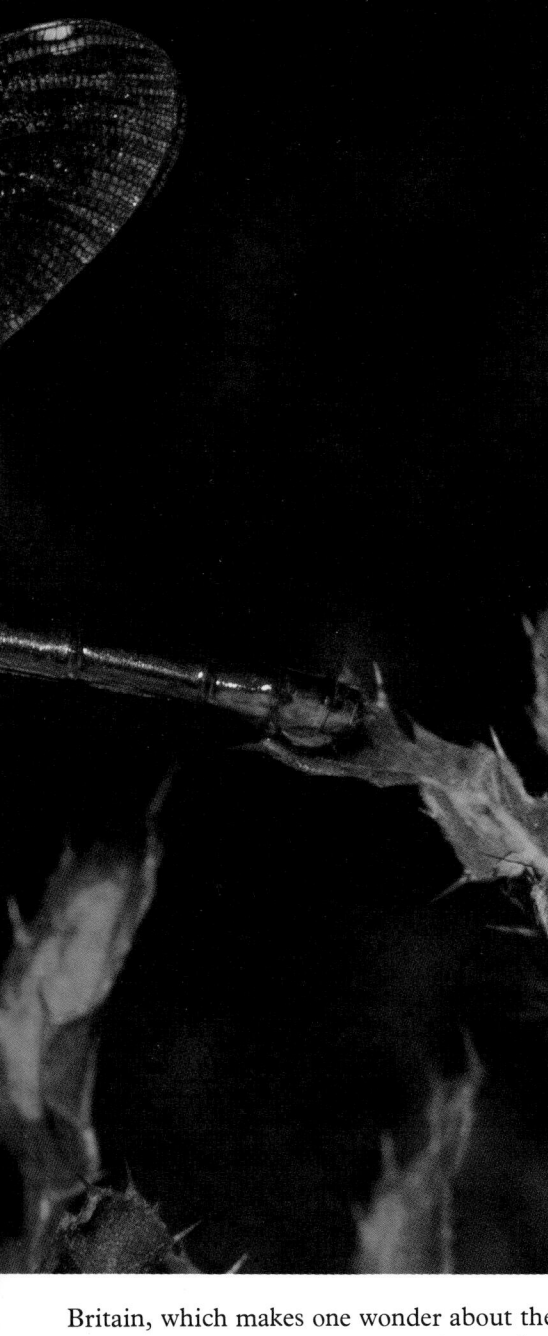

A fine stand of the common fringing grass, reed sweet-grass, *Glyceria maxima*.

Britain, which makes one wonder about the numbers of species we have lost from other not-so-well known insect groups. Many of our dragonflies prefer unpolluted still or very slowly moving waters, but there are a few that like rivers.

'In a weedy section like this, with its mixture of slow and fast flows, you might find the scarce libellula, *Libellula fulva*, although due to the pollution of many of its former sites, it is now very local and only found in southern England. The damselfly that we can see here is one of our commonest species, the splendid agrion, *Agrion splendens*. The ones with the metallic blue colora-tion are the males and the green ones are the females. Both *Agrion virgo* and *A. splendens* prefer fast flowing streams in lowlands which have gravel bottoms. Although many damselflies and dragonflies preferentially frequent still waters, others have an obligate requirement for moving waters. The White-legged Damselfly, *Platycnemis pennipes*, is one such example, being immediately identi-fied by its white tibiae on the middle and back legs. It occurs only in the southern half of England and lays its eggs on floating vegeta-tion. There are also twelve small red or blue damselflies, the common blue damselfly, *Enallagma cyathigerum*, being typical of sluggish lowland rivers with plenty of mar-ginal vegetation. The majority of British large dragonflies prefer to breed in still waters yet the green *Gomphus vulgatissimus*, which is confined to South England and

Wales, breeds in running water, as does the more widely distributed large yellow *Cordulegaster boltonii*.

'Other insects that you might find in well-oxygenated rivers with some gravel on the bottom are the mayflies, Ephemeroptera, and stoneflies, Plecoptera. The former small insects, with their two or three long tails, are probably better known by fishermen than entomologists. The latter, the stoneflies, usually bigger and drab, have two characteristic tails in the nymph stage. The caddisflies, Trichoptera, can be abundant in the lowland rivers. In one such as this, you will find three types of nymphs or larvae: those that live in the gravelly sections and make cases of small pebbles; those that are associated with particular plants, for example, some will use fragments of the water-crowfeet *Ranunculus*; finally, those that are free-swimming and attach themselves to the various surfaces by spinning spider-like nets.

'Feeding on the large numbers of insect larvae and crustaceans that are found in these rivers as well as the plants, are the fish. In a river like this we have all three of the main fish zones. The section we are in at the moment called the trout zone, as it has a fast flow which makes it well-oxygenated and has a shallow, gravelly bottom where the insects, which the trout will feed on, live. Further downstream where the flow is not quite as fast and the river is deeper, is the grayling zone – but because of the influence of the clay I would not consider it to be very good for that species of fish – it is also sometimes called the minnow zone. Then we have the slower weedy stretches, which comprise the coarse fish zone. I would think that you would have dace, roach, gudgeon and chubb along there. There is a fisherman just downstream and if he's had any luck we should find out if I am right.'

We started off down the river and stopped to ask our angler how he was doing. The recent storms had made the water rather turbid but he was in good spirits. He had caught a few dace and knew there were plenty of roach, gudgeon and minnow along that stretch of river. Even a few trout wandered that far down, he confided in us. This nicely confirmed what Nigel had been saying, but Nigel went on to point out that the zone categories are very broad and that in some areas, such as rivers in East Anglia or those flowing through the Fens, the slow coarse section may last for many, many kilometres and there may be no real grayling or trout

zone worth considering. It all depends on the geography of the area.

The bank along this stretch was thick with stinging nettles, *Urtica dioica*, which made walking rather painful but we felt we should not simply ignore them, as every plant has its own story to tell. So we asked Nigel if there was any reason for them to be growing in such profusion.

'It is said that stinging nettles are an indication of rich soils, which is partly true but in many river systems they are usually an indication of disturbance. If you have a bank that is dominated by them it probably means that you are looking at a part of the river that has been dredged and the rich spoil that has been left on the sides has been colonised by the nettles. Once they are established they are very difficult to replace. Here we can see that we are on what I suspect is a recently cut channel which has been created to by-pass the natural meander which you can make out as the depression running across the corner of the field behind us.'

Further along the bank we came to an area that looked far more interesting – a rich variety of plants were fringing the river. We asked Nigel to identify them.

'Here are some nice colourful plants that are highly typical of a lowland river – the great willowherb, *Epilobium hirsutum*; over there is some purple loosestrife, *Lythrum salicaria*, with its spines of purple flowers. Alongside it is a very common plant of these river banks, *Glyceria maxima*, reed sweet-grass, which looks like a smaller version of the common reed. If we look right down at the water's edge, we can see drifting in the current the submerged leaves of bur-reed, *Sparganium*, and next to that are these darker fine-leaved patches. This is a member of the pondweeds, that I mentioned earlier. This particular one is *Potamogeton pectinatus*, fennel pondweed. It is a very common plant in lowland rivers which are nutrient-rich as it can tolerate a high degree of pollution. Where you have clear unpolluted water you have the water buttercups but as soon as the water becomes cloudy then it is nearly always replaced by *P. pectinatus*. It is a very adaptable water plant. Further out in the centre of the river here you can see the bent leaves of the bulrush or clubrush, *Scirpus lacustris*. This species, like the yellow water-lily, *Nuphar lutea*, is very characteristic of rivers on a clay substrate. Most people would think of the waterlily as occurring only in very slow or ponded conditions but, in fact, it will tolerate a fairly fast flow providing it has the right depth and a clayey bottom in which to root. The yellow water-lily has two types of leaves – the floating leathery leaves and thinner cabbage-like leaves which remain under the surface. Therefore, if a boat were to come along and chop off the top leaves, it has always got these crinkly lower ones which can carry on the task of photosynthesising. In this respect it is unlike the white water-lily, *Nymphaea alba*, which has only the large surface leaves. This may well account for the decline of the white water-lily and it is found now only in clean, relatively undisturbed water. This combination of yellow water-lily, bulrush and fennel pondweed with reed sweet-grass and bur-reed on the muddy banks should always tell you that you are in a slow-flowing river with a clay substrate.

'There is a stretch of the River Windrush in Oxfordshire where the river splits into two channels, one goes over a fairly gravelly substrate while the other is clay and although there is no difference in the water between the two channels, their floras are totally different. In the one with the gravel it is almost totally dominated by the brook crowfoot, *Ranunculus calcareus*, and the river crowfoot, *R. fluitans*. In the other stream,

where you have the clay, it is all bur-reeds, yellow water-lily and bulrush, simply because the habitat is different.

'Where the substrate of the river is very loose and silty you will only find those plants, such as the starworts, Canadian pondweeds, *Elodea*, and hornwort, *Ceratophyllum*, which are shallowly rooted and able to cope with the unstable floor of the river. Plants, such as the water-lilies, need to root into a more solid substrate and if there is a layer of silt they either have to be able to grow up through

Above **The Blythe, earlier in the year, with a white sheet of** *Ranunculus* **flowers across its surface.**

Below **A lone bullock drinking at the river's edge. Cattle trampled areas can provide an important habitat for many riverside annual plants.**

Above **A large hoverfly feeding on the terminal flower-head of water-mint,** *Mentha aquatica.*

Centre top **A yellow waterside carpet of great yellow-cress,** *Rorippa amphibia,* **typically growing half-in and half-out of the water.**

this, which presents problems for new shoots as this layer is often very anaerobic in rich systems, or they are unable to put down roots. It might appear to be alright for one season but when the plant dies back for the winter then tries to put out its tender new shoots in the spring, they are just smothered. This problem is not always appreciated by river engineers who widen a river course to cope with periods of flood and find that during the summer season the low flow in the river is so dissipated that silt, which would otherwise be moved on, is deposited across the river bed. A further problem is that any plants that root in this silt are washed away if the river flow suddenly increases.'

We now made our way further along the riverside. Ahead of us was a herd of cattle which were drinking from the river. This part was obviously a regular drinking spot as

the bank had been broken down and the edge was well trampled. This provided an interesting series of habitats for waterside plants, as Nigel explained.

'Cattle trampling can actually increase the diversity of the plant community, as it keeps the surface disturbed and open, allowing annual species to get a hold, whereas normally they would have to compete for space with the tall growing perennials that crowd the bank. Here we have a rich mixture of both low-growing perennials and some annuals that are straggling from the muddy flat bottom of the bank into the river. First, there is water mint, *Mentha aquatica*, which is on the landward side where it doesn't become too wet. Then we have another member of the labiates – gypsywort, *Lycopus europaeus*, which has deeply toothed leaves and tiny white flowers borne in whorls around the stem. Right on the water margin we have two plants – water forget-me-not, *Myosotis scorpioides* and great yellow-cress *Rorippa amphibia*. This last plant is appropriately named as it is truly amphibious and will grow both on land and in water. At certain times of the year it will produce a fantastic yellow carpet of flowers along the bank. There are actually four types of watercress, two others of which have yellow flowers: *Rorippa sylvestris*, which is rarely found in lowland rivers because it prefers to grow amongst loose

pebbles in unstable situations; *R. palustris*, which is more frequent, growing in marshy and muddy cattle-trampled areas such as this. In chalky and base-rich rivers is the well-known white-flowered watercress, *Rorippa nasturtium-aquaticum*, which grows best in base-rich mud.

'You will have noticed that a lot of these plants have the word *aquaticum, palustris* or *fluviatilis* in them which simply denotes that they are found in either aquatic, marshy or river habitats. For example, here we have *Alisma plantago-aquatica*, which is the common water-plantain. It has these leaves that look like those of a plantain but it belongs, in fact, to its own separate family.

'Just here is an interesting plant. This is trifid bur-marigold, *Bidens tripartita*, which is a fairly uncommon annual of these trampled sites. The term "trifid" implies nothing sinister but refers to the trifoliate leaves. When the fruits ripen they have little barbed spines on top which presumably fasten on to animals that brush past to drink at the river, so dispersing the seeds. The more common, nodding bur-marigold, *Bidens cernua*, occurs in similar situations but has long undivided leaves and a drooping flower-head. We also have here two other annual plants that grow well in these more open situations. First, here is toadrush, *Juncus bufonius*, which has small pale green flowers and narrow grooved

leaves. Then, next to it, we have a very small specimen of celery-leaved buttercup, *Ranunculus sceleratus*. It has these little yellow flowers and a good specimen can grow to over ninety centimetres (three feet) with a thick stalk and large lobed leaves.'

The opposite side of the river had a low vertical clay bank with many holes arranged along it above the water level. We asked Nigel what might have made these.

'Vertical banks are often important for nesting birds. If they are high enough you might find sand martins nesting in them. Where the nesting holes have been unavoidably destroyed it has been found that the martins will use sections of pipe pushed into the bank. But these holes are too low. I would guess that they were made by water voles when the bank was slightly lower and it has since been eroded away to expose them to

Above **A small fly resting on the flowering spike of water forget-me-not,** *Myosotis scorpioides*, **a common plant of river-sides throughout Britain.**

Above **One of the more local riverside annuals is this member of the composites – trifid bur-marigold,** *Bidens tripartita.*

Above right **The oddly-named skullcap,** *Scutellaria galericulata,* **with its pairs of labiate flowers along one-side of the stem. It is a fairly common plant of riversides and marshes.**

view. Water voles like well vegetated banks to hide amongst. This enchanting rodent is often unkindly called the water rat but it doesn't look like a rat at all, having a much more rounded head with small ears and a shorter tail. They are almost exclusively vegetarians, unlike the other common water-side mammal – the water shrew. This is a voracious carnivore, like all the shrews, and will eat anything from worms and small insects to tadples and small fish. It is a good swimmer, like the water vole, and can be easily recognised by its small size and contrasting black upperparts and lighter underside.

'A much rarer animal of our lowland rivers is the otter. We do not know why so many otters have disappeared from lowland England. There are various theories which include factors such as the extra disturbance of very rapid and intense land drainage work; the effect of pesticide run-off from neigh-bouring farmland and generally increasing disturbance along our rivers. It is almost certainly due to a combination of factors. There is so much sympathy for the otter that certain areas, which have a suitable range of habitats, have been designated as otter "havens". In some places in East Anglia they are transplanting otters back into the wild but this has to be carried out very carefully as the otter has a mammoth range, for instance, a dog otter will travel anything from ten to twenty kilometres (six to twelve miles) in a couple of days.

'Looking again at the range of plants that we have along the bank here, we have a curiously named plant – the skullcap, *Scutellaria galericulata.* The name refers to the shape of the calyx which is supposed to resemble a leather helmet worn by roman soldiers, called a *galerum.* This is a fairly common plant of banksides, though its numbers have suffered from the drainage of many of its former marshy haunts. It has small blue tube-like flowers. Here is another plant with blue flowers, though admittedly of a very different shape – *Veronica beccabunga,* brooklime. This member of the figwort family is very common along our lowland rivers. It has blunt fleshy leaves and a tall spike of deep blue flowers. There are two other speedwells that are found along rivers – blue water-speedwell *Veronica anagallis-aquatica,* and pink water-speedwell *V.*

catenata. The former is found typically in chalkstreams, and is fairly widely distributed, even being found in the oolite streams of Derbyshire. *Veronica catenata* has pink flowers and, although it is more catholic in its requirements, it is less common, probably because it is not as robust as *V. anagallis-aquatica*.

'Growing in the water's edge we have a rather striking plant – arrowhead, *Sagittaria sagittifolia*, with these white three-petalled flowers and very imposing aerial leaves shaped appropriately like large arrowheads. Interestingly, the arrowhead has three different types of leaves: these arrow-shaped ones, then floating on the surface we can see some, which more closely resemble the leaf of a broad-leaved pondweed and finally, submerged in the river, are more strap-shaped leaves, which are very undulant. This one is in a bay where the water is rather sluggish so the aerial leaves are dominant but where there is more flow the linear, submerged leaves become dominant and you could mistake it for a different plant – in fact, they look rather like the leaves of the bur-reed, *Sparganium emersum*. A similar phenomenon occurs with bulrush, *Scirpus lacustris*. Most people know this plant as a tall emergent plant, forming dense bankside stands, but it can grow in a good current and it will then have long submerged leaves and no flowers.

'Around the base of the burreeds in this bay are some duckweeds which, if you remember, are free-floating plants. This is the common duckweed, *Lemna minor*, which consists of a simple flat frond or thallus, only two to four millimetres across, and a single long root hanging down below. In very sluggish rivers you might find fat duckweed, *Lemna gibba*, which has an air sac underneath the thallus. In dykes you will find a larger duckweed with a tuft of roots, *Lemna polyrhiza*. It is not really a river species but you might discover it where a slow-flowing river is fed by a dyke filled with this plant. Both *Lemna gibba* and *L. polyrhiza* produce special winter forms which sink to the bottom of the river or pond to lie dormant until the spring when they bud-off new free-floating forms. There is a fourth species that you might find – this is the ivy-leaved duckweed, *Lemna trisula*. This species tends to remain predominantly submerged sometimes carpeting the bottoms of rivers which have clear water. It has strange translucent fronds that usually form branched colonies

that look like groups of tiny ivy leaves.'

As we were peering down at these unusual little plants, a mute swan glided serenely past looking at us as if to say: what's so special about my dinner? This was the first swan we had seen that day and we asked Nigel if they were a common sight along the river.

'They seem to be alright along this stretch but unfortunately our population of swans is generally suffering rather badly from lead poisoning. Although, in many things, the fishermen are our allies it does seem that one of the main causes of this is discarded lead shot left by anglers. Like many birds, swans habitually swallow pellets to aid their digestion and, in doing so, take in lead shot from the banks and river bottoms. This then builds up, eventually affecting their nervous system. Sometimes it affects their gut muscles so that they starve to death even though their crops may be full. Disturbance and even vandalism has not helped matters

A lone mute swan framed by the bankside trees in the evening light. Our population of swans has been declining and recent reports indicate that lead poisoning is the main reason for this.

Along deeper, slow-flowing sections of the river the bizarrely shaped leaves of the arrowhead, *Sagittaria sagittifolia* can be seen. This plant has, in fact, three distinct types of leaves – submerged strap-like leaves, floating pondweed-like leaves and the arrow-shaped aerial-leaves.

and I'm afraid in some areas already the population of mute swans has decreased dramatically. A simple thing to help prevent this that any angler can do, is to make sure that the lead weights and split shot they use are not discarded or left on the banks where the swans can reach them. It is important that people realise that lead shot on the bed of the river or discarded in the thick vegetation on the bank are both as potentially lethal to the swan.

'Other birds that you might hope to see along here are moorhens and coots, the latter preferring slower flowing rivers. I know that there are kingfishers along here but that is a bird you have to have a great deal of patience to see.'

A little further down the river was a classic meander in the river which clearly showed that the river was still making its own course along this part of its length. Nigel took up the story.

'Here we can see a meander with its typical eroding and depositing sections. What is happening is that on the opposite bank the flow is cutting deeply into the edge and making a vertical bank, and on this side you have the deposition of any gravels that have come downstream. If this was a more silty river, this bank would have a deposit of silt,

probably colonised by reed sweet-grass. So here we can say that the river is being allowed to do its own thing. Often when river engineers "improve" a stretch of river they will try and cut out the meanders, as we saw further up the river. Even if they leave them in, they will frequently dredge so that the typical shelving profile is taken out. If we look just down stream of the meander we can see that after the deep pool section there is an area where the river is much shallower causing the surface to look fairly turbulent – this is called the gravelly riffle section which will be well oxygenated and will be colonised by *Ranunculus*, in this case, the river crow-foot, *R. fluitans*. If the meander continues to develop naturally the curve will become more and more pronounced until the river will break through the thin strip of land between the two ends of the curve, cutting off the original bend to form a new channel and leaving an ox-bow lake. The pond-like habitat that you get in these cut-off ox-bows can be very rich and, if it is still periodically replenished from the river during flood times, the water will remain relatively fresh which is better for the aquatic life.

'We are now walking beside a more gravelly section which in early summer is white with the crowfoot flowers. Having just seen a natural meandering section we now have a very straight stretch which has been made by man, almost certainly to divert the river away from the railway, which runs alongside just here. It still has a gravelly bottom but soon we will be coming to a section where the river water has been ponded back because of a mill. One further point we should mention here is the algal growths that occur in these rivers. When the water becomes nutrient rich, possibly as a result of sewage input or the drainage of heavily fertilized farmland, the algae tend to do better competitively than the higher plants and can become a dominant feature of the river vegetation at certain times of the year. In this stony section with its reasonable flow you will find a species called cotweed or blanketweed, *Cladophora glomerata*. It can become so prevalent that it smothers the other plants. Another filamentous alga that you find in lowland rivers is *Enteromorpha*, which is related to the blanketweed you find on the coasts. It is a very characteristic algae, forming long green tubes, often more than a metre long. In East Anglia a bright green carpeting alga, which has an almost pelt-like look about it, is *Vaucheria*. Large "blooms" of these species are looked upon as a sign of secondary pollution and on decompsition can severely deplete the oxygen content of the river water. In undisturbed rivers subjected to an increase in nutrients, it is usual that the larger plants will rebutt the damaging growth of the algae due to two main factors: firstly, they already occupy some of the prime sites in the river; secondly, they often produce substances, which they release into the water, that subdue algae growth. In a recently dredged river such a balance is lacking and these are the areas where algae problems are greatest.

'Here we are just a little further along and the river has changed character yet again. There are bur-reeds growing out into the water and patches of water-lilies that tell us we are back on a clay substrate. Growing here are the two species of *Elodea* waterweeds. The Canadian waterweed, *Elodea canadensis*, was introduced in British rivers over 100 years ago (1847) and rapidly spread throughout Britain's waterways to such an extent that it became a threat to navigation, clogging up canals and rivers. In fact, a Minister was appointed to control the problem in the 1860s. Interestingly, this plant was spreading vegetatively as there was only the female plant present. The vigour that this

coloniser had in the nineteenth century appears to have gone and it seems to have reached a steady state. However, another member of the same genus, *Elodea nuttallii*, was first noticed in East Anglia in 1975 and has since spread to many parts of England and may well be replacing *E. canadensis*. It is particularly prevalent in deep, nutrient rich waters with sluggish flows.

'Further along here the river banks appear very different to the fairly open sections that we have just seen. This is because as we approach the mill the water has been held back and consequently there has been a lot of deposition along the banks which are now much more silty and muddy. Colonising these sections are large stands of reedmace,

The linear nature of our riverside habitats mean that introduced plants can sometimes spread rapidly. The case of the aquatic Canadian waterweed has been well documented. The Himalayan balsam, *Impatiens glandulifera*, **is another case. This plant is another introduction, this time from North America, orange balsam,** *Impatiens capensis.*

83

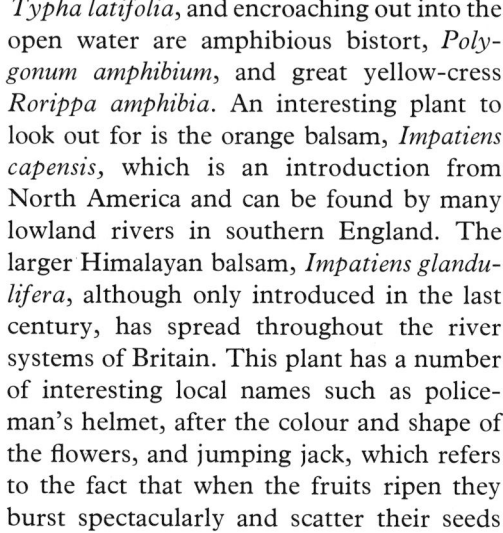

As the river approached the mill the water became increasingly ponded-back resulting in a thick band of reedmace and bur-reed fringing the river and large amounts of water-lily growing over the open surface.

Centre **A classic example of a meander showing the deposition on the inside of the bend and the undercut bank on the outside. Just down from the meander the water surface is broken by a riffle. All too often when rivers are 'improved' features such as these, with all the varied habitats they provide, are simply removed by dredging or re-channelling.**

Typha latifolia, and encroaching out into the open water are amphibious bistort, *Polygonum amphibium*, and great yellow-cress *Rorippa amphibia*. An interesting plant to look out for is the orange balsam, *Impatiens capensis*, which is an introduction from North America and can be found by many lowland rivers in southern England. The larger Himalayan balsam, *Impatiens glandulifera*, although only introduced in the last century, has spread throughout the river systems of Britain. This plant has a number of interesting local names such as policeman's helmet, after the colour and shape of the flowers, and jumping jack, which refers to the fact that when the fruits ripen they burst spectacularly and scatter their seeds along the bank and into the water.'

We then made our way up onto a road bridge that crossed the river. Here we had a view of the mill laid or leet, which formerly would have taken the water from the river to power the mill wheel. It was obvious from the almost pond-like nature of the water that it had long been in disuse. The 'pond' was fringed with a band of reeds, laced with flowering plants and the surface of the water was broken by patches of yellow water-lilies. A perfect habitat for dragonflies and other water insects. Running off from this still area of water was a narrower channel which eventually joined the main course of the river on the other side of the mill. Here we were confronted with a very different spectacle to the quiet scene we had just been surveying. Nigel explained what was happening.

'Here we can see some river dredging work in progress. The idea is to attempt to recreate a natural channel at about three-quarters of a metre (two feet) below the present river bottom. The usual management practice is to

scrape off the top soil adjacent to the river and put it to one side, then scrape out the river bed and use that to raise the level of the bank, covering this with the original top soil. This means that the land can still be utilised by the farmer. It has a dual purpose as it stops the floodwater spilling into the fields and provides an accessible area for them to dump the spoil. You can see that they have found a mix of clays and gravel in the river bed. It is a very expensive operation and one often questions the overall cost benefits of such schemes. The purpose of this is to allow the farmer to change from using this adjoining land as pasture to arable. As pasture the land can often be helped by periodic flooding from the river as the deposited silts will enrich the land. The fishing fraternity obviously are anxious to maintain the quality and type of river as it is above the mill and, in that sense, are important allies of the conservationists when it comes to objecting to the proposal of such schemes'.

We had certainly appreciated our walk along this stretch of river. The feature which had been most surprising was the variety of habitats that these more natural lowland rivers could provide – the open fast flowing sections with their plants twisting and turning in the current; the more tranquil slow sections with the insects darting from flower to flower along the bank; the shade of the riverside trees and the unexpected view around the meander. It would be a sorry day if this was all reduced to the bland uniformity of a simple 'drainage channel'. Happily, there are still many hundreds of miles of walks by our lowland rivers that provide a refreshing glimpse of our natural heritage in all its forms and fortunately, there are people like Nigel keeping a careful watch on them.

Dredging in progress. The crane with its giant bucket has been taking material from the bed of the river and dumping it on the adjacent field. This highly expensive operation is essentially to help the farmer grow more cereal crops by stopping the river from flooding in winter.

A chalk stream walk

A chalk stream is a delight at any season with its clear water, 'cool' in summer, 'warm' in winter, overflowing its banks in due seasons and bursting with life of every kind. It is one of the perfect features of the English countryside. Fortunately some are still managed as part of the rural economy, as they have been since the middle ages and before. Great drifts of water-buttercup and starwort, the vibrant green masses of their stems and leaves acting like a series of coffer dams, ponding the water back and providing hiding places for plump brown trout. The fish themselves rising to a multitude of 'flies' both natural and man-made. Most important of all, is the water itself, charged with calcium, which helps to build the carapaces of multimillions of freshwater shrimps and the shells of water snails, and clear, so clear, that nothing stays hidden from view. The work of the Freshwater Biological Association's River Laboratory is adding all the time to our knowledge of our chalk streams and helping to ensure their future.

A tranquil scene along one of our rich southern chalk streams. Towering willows and fragrant meadowsweet frame the picture.

Information

The chalk streams, of all the lowland river types, are probably the most rewarding both for plants and animals. Their nutrient-rich waters and stable regime means that they are an ideal habitat for a rich diversity of plants and attendant invertebrates. Because the water is mostly derived from springs, it has little solid material in suspension and consequently the water is usually clear. Also, the stream bed has little silt and a gravelly bottom which provides reasonable anchorage for plants. This chapter describes a walk along a chalk stream in early summer when the system is probably at its most productive.

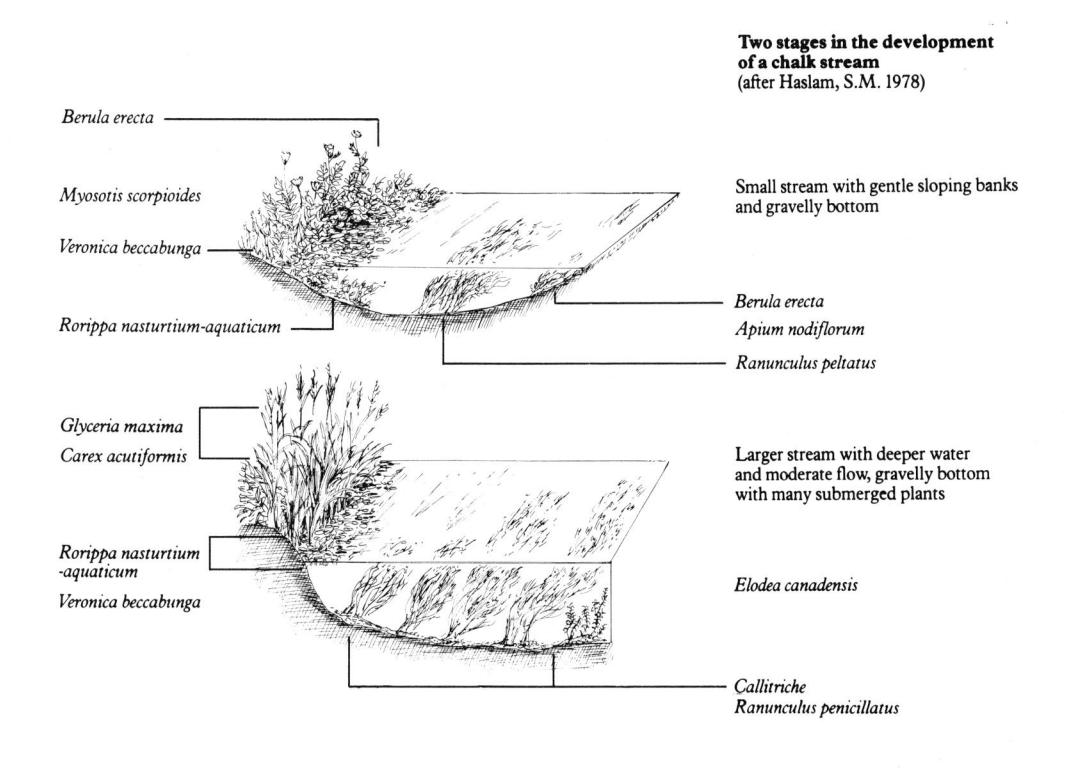

Two stages in the development of a chalk stream
(after Haslam, S.M. 1978)

Berula erecta

Myosotis scorpioides

Veronica beccabunga

Rorippa nasturtium-aquaticum

Small stream with gentle sloping banks and gravelly bottom

Berula erecta
Apium nodiflorum

Ranunculus peltatus

Glyceria maxima
Carex acutiformis

Rorippa nasturtium -aquaticum

Veronica beccabunga

Larger stream with deeper water and moderate flow, gravelly bottom with many submerged plants

Elodea canadensis

Callitriche
Ranunculus penicillatus

Sites

The most famous chalk streams arise in the large outcrop of chalk forming Salisbury Plain in Wiltshire and the neighbouring Hampshire Downs. They drain southwards across the coastal Tertiary deposits in the Hampshire Basin before reaching the Solent. The most well-known of these rivers are the **Avon, Test, Itchen** and **Meon.** Most of these rivers are prize fishing waters and consequently public access is generally restricted. However, if you use a good Ordnance Survey map which shows the public rights of way it is usually possible to find a stretch with a footpath.

Canals

Most of our chalk streams have been managed over the centuries either to provide power to drive watermills or to irrigate the neighbouring water meadows. One type of lowland 'river' that has been created by man is the navigation canal. These canals formed a vast network of waterways during the first half of the nineteenth century and although, with the advent of the railways and roads, they have fallen into disuse, many have been kept-up by enthusiastic canal societies. The water in these canals is generally slow-flowing and the flora and fauna is rather similar to that which you might find in an enormous linear pond or very slow-flowing river. The great bonus with them is that the towpaths, that were used originally by horses pulling the barges, have usually been turned into public rights of way. These canals can be very rewarding places to explore for plants and animals. The **Grand Union Canal** between Kilby Bridge and Market Harborough is managed by the Leicestershire and Rutland Trust for Nature Conservation in agreement with British Waterways and has a varied aquatic and waterside flora and fauna. Some other canals that are worth visiting are the **Brecon and Abergavenny Canal,** the **Kennet and Avon Canal,** the **Basingstoke Canal** and stretches of the **Leeds-Liverpool Canal.**

Plants and animals to look out for in chalk streams:

Plants:

river crowfoot, *Ranunculus peltatus*
   Ranunculus penicillatus var. calcareus
water-cress, *Rorippa nasturtium-aquaticum*
fool's water-cress, *Apium nodiflorum*
lesser water-parsnip, *Berula erecta*
river water-dropwort, *Oenanthe fluviatilis*
blue water-speedwell, *Veronica anagallis-*
   aquatica
brooklime, *Veronica beccabunga*
water forget-me-not, *Myosotis scorpioides*
water mint, *Mentha aquatica*
branched bur-reed, *Sparganium erectum*
starworts, *Callitriche*

Depending on the management of the banks
the surrounding vegetation can be every bit as
profuse and might include the following:
yellow flag, *Iris pseudacorus*
meadowsweet, *Filipendula ulmaria*
water avens, *Geum rivale*
purple loosestrife, *Lythrum salicara*
yellow loosestrife, *Lysimachia vulgaris*
great willowherb, *Epilobium hirsutum*
water figwort, *Scrophularia aquatica*
common comfrey, *Symphytum officinale*
hemp agrimony, *Eupatorium cannabinum*
common reed, *Phragmites communis*
reed sweet-grass, *Glyceria maxima*
greater tussock sedge, *Carex paniculata*
lesser pond sedge, *Carex acutiformis*

Fish

Chalk streams provide some of the finest
fishing in Britain and are particularly
renowned for their salmon and trout fishing.

salmon, *Salmo salar*
brown trout, *Salmo tutta*
rainbow trout, *Salmo gairdneri*
grayling, *Thymallus thymallus*
bullhead, *Cottus gobio*
minnow, *Phoxinus phoxinus*
stone loach, *Neomacheilus barbatulus*

Invertebrates

Chalk streams have extremely large
populations of invertebrates. The *Gammarus*
shrimp is probably the most numerous species
but there are also important populations of
freshwater operculate snails and insect larvae
such as caddis flies and mayflies. The
following list just illustrates a few examples
that might be found.

freshwater shrimp, *Gammarus*
freshwater crayfish, *Astacus pallipes*
bladder snail, *Physa fontinalis*
nerite, *Theodoxus fluviatilis*
freshwater winkle, *Viviparus viviparus*
banded agrion, *Agrion splendens*
white-legged damselfly, *Platycnemis pennipes*
mayfly, *Ephemera danica*
mayfly, *Baetis*
stonefly, *Nemoura cinerea*
caddisfly, *Brachycentrus subnubilus*

Birds

little grebe, *Tachybaptus ruficollis*
heron, *Ardea cinerea*
mute swan, *Cygnus olor*
mallard, *Anas platyrhynchos*
tufted duck, *Aythya fuligula*
moorhen, *Gallinula chloropus*
coot, *Fulica atra*
grey wagtail, *Motacilla cinerea*
kingfisher, *Alcedo atthis*

Mammals

water shrew, *Neomys fodiens*
water vole, *Arvicola terrestris*
otter, *Lutra lutra*
American mink, *Mustela vison*

A chalk stream walk

with Alasdair Berrie

The rich tapestry of water plants clearly underlines the quality of the water. Here you can see a mix of water crowfoot, *Ranunculus*, showing as darker green patches, with lesser water-parsnip, *Berula erecta*, and starwort, *Callitriche*.

Opposite **The exquisite white-petalled flowers of *Ranunculus* standing out above the river surface.**

Sometimes when you go for a walk in the countryside you feel as though the whole place is humming with life. Everywhere you look there is something exciting to see – a butterfly that has suddenly landed right at your feet, a bright patch of flowers that seem unfamiliar, even a glimpse of a kingfisher as it flashes past you on its way upstream. It was just such a day as this when we went for a walk along a stretch of chalk stream in rural Hampshire. We met our guide, Dr Alasdair Berrie, head of the Freshwater Biological Association River Laboratory in Dorset, beside one of the major roads that crosses through central Hampshire carrying traffic from London to the coastal conurbations. At first glance the countryside appeared to be of the kind through which the motorist so often passes with blinkered eyes – nothing special, nothing that catches the attention. Yet within a few yards of the busy main road we were surveying a landscape which has enthralled naturalists, walkers and fishermen for hundreds of years. The chalk streams of southern England have long been praised for the excellent quality of their fishing. This degree of excellence is closely allied with the wealth of animal and plant life that abounds in this rich habitat. It was difficult to know what to look at first but Alasdair went straight to the life-blood of the landscape and told us about the importance of the type of river.

'We've come to one of Hampshire's famous chalk streams, the River Itchen, and are now looking at a piece of very traditional English countryside – a lowland river, fringed with reeds and overhung at intervals with willows. However, as this is chalk country and the rock in the catchment area of this river is also chalk we should appreciate what influence this has on the hydrology of the river itself. The striking fact about water in a chalky area is that because of the porosity of the rock, the water drains down through the chalk rather than running off the land surface. This means that you have great underground reservoirs from which water moves laterally through the rock to come out of the ground at springs which combine to form the headwaters of the river. The whole system is rather different from most rivers, certainly from those in upland areas where the main source of water is from surface run-off. There are four main factors that result from this that make these rivers such ideal environments for animal and plant life.

'Firstly, the flow of the chalk streams is very stable because it takes time for the water

to percolate through the chalk, both in its journey down to the underground reservoirs and in the lateral flow to the springs. So when you have heavy rainfall it does not immediately produce an outrush of water into the river. Chalk streams very rarely suffer from spates which could uproot much of the plant life.

'Secondly, these underground reservoirs tend to have a very constant temperature (about 11°C) which reflects the temperature of the rock. So the temperature of the spring water is more or less constant throughout the year. In the winter the water that is coming into the river is relatively warm compared with that which is already in the river and it becomes progressively cool as it goes downstream. Conversely, in summer, the water coming into the river is relatively cool. The consequence of this is that the temperature range in a chalk stream is much less than in other rivers. They seldom fall below about 5°C in the winter and certainly do not freeze. In the summer they seldom rise above about 17°C.

'Both these factors tend to break down if you go to the lower reaches of many streams because, as on this river, they may no longer be on chalk and will be receiving surface water from the surrounding land. However, on a genuine chalk stream you do have this relative stability of flow and temperature. This makes the rivers themselves very good places for plant growth and if we look out across here we can see that the bed of the river is a lattice of different coloured greens. The fact that we can see all these plants brings me to the third influence of the chalk.

'Typically chalk streams are very clear because the water has been gradually filtered as it has passed through the chalk and emerges as pure spring water. Conversely, where you have rivers which are fed by surface water, this has picked up soil particles which make the water cloudy. As you go to the lower reaches of a chalk stream you may begin to find that the water is less clear. Towns or cities will be releasing their sewage effluent into the river and, even though it is probably extremely well-treated, it will inevitably contain small particles suspended in the water. So although the river does not suffer any serious pollution it does lose clarity, which could well affect the plant growth in the deeper sections as the light will not get through to enable plants to photosynthesise.

'The fourth major effect of the chalk is that the water is able to dissolve considerable quantities of nitrates, phosphates and potassium, and comes out of the underground reservoirs rich in plant nutrients. Just the things that you put on your plants in the garden. If you have a river which is running off a hard rock, such as in some mountainous areas, then very little material from the rock goes into solution. But the plants rooted in the riverbed here are being bathed in a continual flow of ''fertilizer''. So there are good reasons why this environment is so rich.

'Well, we have already commented on this lovely patchwork of greens in front of us. If we look a little closer we can see that they consist of three main species. There's one with very dark green elongated strands out in the middle of the river. That is a water crowfoot, *Ranunculus*. This group of plants is rather complex containing many species, some of which grow in fairly stagnant or slow-moving waters, whilst others such as this one prefer faster flowing streams – this is probably *Ranunculus penicillatus*. They are an aquatic relative of the familiar buttercups and have white buttercup-like flowers.

'Interestingly, they flower earliest upstream. We are not sure why this is, whether

In reaches where the flow is slower and silt has been able to accumulate on the river bed starworts, *Callitriche*, can become dominant. Here is a bank of starwort, with its typical almost sculptured shape, next to some darker strands of *Ranunculus*.

there is a genetic difference in the plants or whether it is simply that it takes longer for the plant to grow and reach the surface as the river becomes deeper. There is some experimental evidence that if you move a plant from upstream and keep it in a downstream situation it will still flower earlier. It is probably a complex mixture of heredity and environmental factors, as is often the case. If the river is shallow enough and the growth of *Ranunculus* is large enough you can see great masses of these white flowers across the surface in summer looking quite spectacular. Strangely, although they do flower nearly all the new *Ranunculus* beds are formed by vegetative propogation and not by seed. In other words, what usually happens is that a little part of the plant breaks off and is carried off downstream and may eventually root itself if it becomes trapped where suitable conditions occur.

'The bright green beds of weeds that have an almost sculptured look to them are starworts, *Calltriche*. These are plants like the water-crowfoots that root themselves into the bed of the river and can become dominant where there is a build up of silt, so they often indicate areas where the flow has slowed allowing material carried in suspension to be deposited. The third type of submerged plant which is common in the river here with the distinct branched leaves is lesser water-parsnip, *Berula erecta*, a member of the parsley family. These three plants occur right across the river but along the banks you have dense stands of emergent plants which are excellent in providing cover for all types of animals.'

Across the river we could see a family of coots foraging amongst the fringing plants with the newly hatched chicks with their red heads, paddling after the parent birds. Further up the river a pair of tufted ducks were swimming – the handsome black and white plumage of the drake contrasting markedly with the subdued browns of the duck. They probably had a nest somewhere amongst the dense riverside herbage and had been disturbed by the fishermen patrolling the river. Unlike the familiar mallards, which feed from the surface, tufted ducks are expert divers and as we watched them they frequently disappeared from view almost certainly to feed on the rich growth of plants on the river bottom.

Alasdair pointed out another fact about the luxurious river vegetation.

'The profusion of water plants has an additional hidden benefit to the river animal life, for whilst they are all photosynthesising they release oxygen into the water and although they will use some themselves for respiration, the net benefit to the river is considerable. The fishermen are particularly pleased to see these plants doing well as not only do they indicate that the water quality itself is good but they are providing important shelter and a good supply of oxygen for the trout. However, if you get too much weed cover it becomes almost impossible to fish

The dense bank side vegetation provides a superb habitat for water birds such as this moorhen.

The freshwater shrimp, *Gammarus*, is found in enormous numbers amongst the stones in the river bed and is the main source of food for the chalk river fish.

hundreds of miniature dams built across it. As the water is ponded back it can spill out onto the surrounding land causing summer flooding, even though there may not be a higher volume of water in the river.

'For centuries these chalk streams have been recognised as perhaps the most excellent trout fishing rivers in the country. This is partly because of the luxuriant plant growth which we can see but also because of the vast amount of insect life that lives in the river bed and amongst the weeds. This provides a good supply of food and the relatively narrow range of temperature variation combines to give almost ideal conditions for the growth of trout. Isaac Walton himself, who died exactly 300 years ago, is buried in Winchester and was a great enthusiast for chalk stream trout.

'The best place to look for trout is just on the edge of a weed bed. You might see one exploring out or breaking the surface to snap up a mayfly. A good time to watch them surface feeding is in the morning or evening when the insects are swarming over the water. If the light is reflecting off the surface of the water, a pair of polarised sunglasses will reduce the glare and will enable you to see the trout darting along just above the river bed.'

As many of the insects are more active at dusk and tend to hide up during the day, we turned our attention from the open water to the fringing plants and before long we spotted a brownish insect rather like a

effectively and therefore the weed beds are cut from time to time. The local water authority will also cut the weeds back to prevent flooding. In chalk streams the flow of water is usually greatest in the spring and lowest in the autumn. However, during spring and early summer you have a great increase in the volume of the weed beds and this can produce a damming back effect on the whole river. It is just as if you had

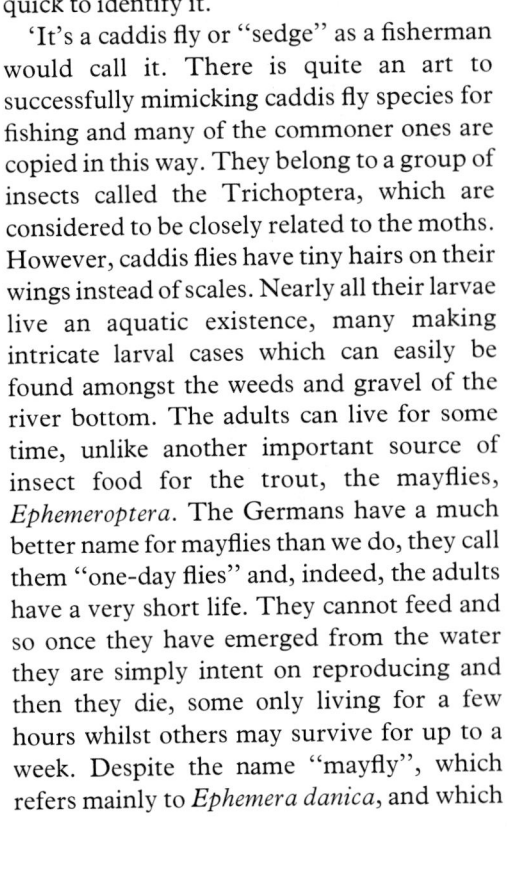

narrow moth resting on a leaf. Alasdair was quick to identify it.

'It's a caddis fly or "sedge" as a fisherman would call it. There is quite an art to successfully mimicking caddis fly species for fishing and many of the commoner ones are copied in this way. They belong to a group of insects called the Trichoptera, which are considered to be closely related to the moths. However, caddis flies have tiny hairs on their wings instead of scales. Nearly all their larvae live an aquatic existence, many making intricate larval cases which can easily be found amongst the weeds and gravel of the river bottom. The adults can live for some time, unlike another important source of insect food for the trout, the mayflies, *Ephemeroptera*. The Germans have a much better name for mayflies than we do, they call them "one-day flies" and, indeed, the adults have a very short life. They cannot feed and so once they have emerged from the water they are simply intent on reproducing and then they die, some only living for a few hours whilst others may survive for up to a week. Despite the name "mayfly", which refers mainly to *Ephemera danica*, and which

does emerge in late May, various other species can be seen on the wing throughout the summer. An interesting fact about these primitive creatures is that they are the only insects that moult after they are able to fly. The newly emerged "sub-adults" are called "duns" by the anglers, as they are covered in fine hairs. After a few hours they moult for the last time to become the glistening adults that the fishermen call "spinners". The nymphs live for much longer periods, some taking two years to reach the adult stage.

'The amount of invertebrate life in the river is quite amazing. The most common animal on the bottom in these rivers is usually the freshwater shrimp, *Gammarus*, which grows to about three-quarters of an inch (18 mm) in length. They are present in the gravel and weed beds in enormous numbers – anything up to about 20000 to a square metre. If you disturbed almost any small stone down there one or more of these little shrimps would come out. The next most common animals would probably be one of the insects, either mayfly or midge larvae. There are very large populations of midges, fortunately mostly ones which don't

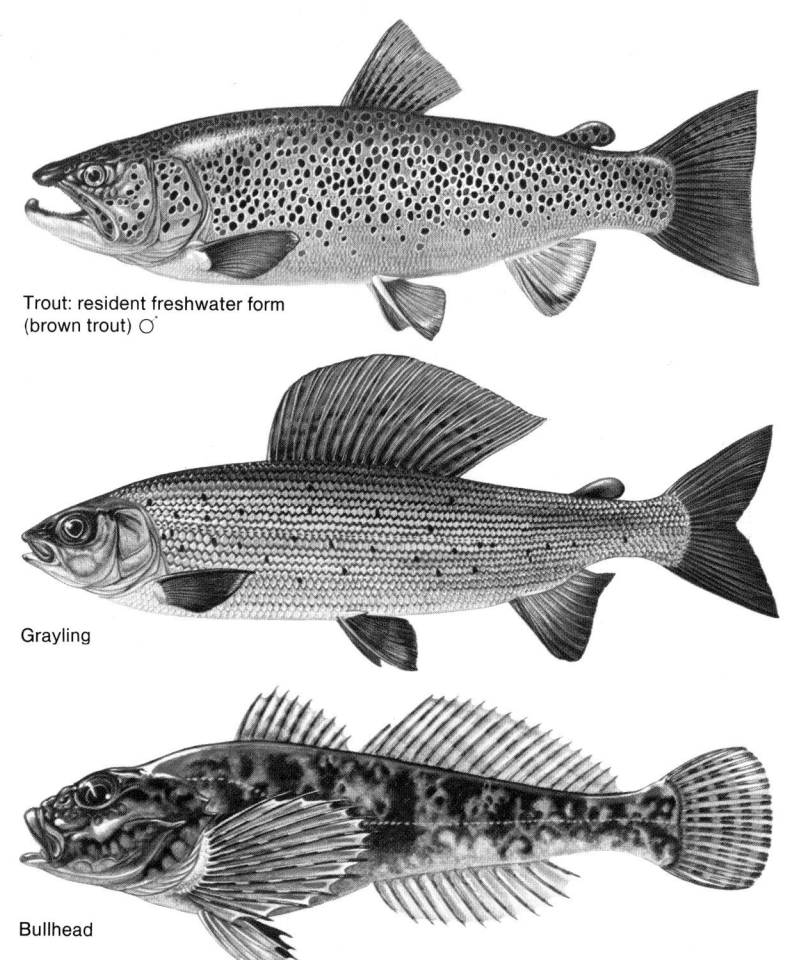

Trout: resident freshwater form (brown trout) ♂

Grayling

Bullhead

Three typical fish of chalk streams: brown trout (top), grayling (centre) and bullhead (bottom).

Opposite top **This little mayfly is probably** *Baetis rhodani*. **It clearly shows the long 'tails', the large front wings and much smaller hind wings. The wings are always held vertically as they are unable to fold them back.**

Opposite below **When most people talk of the mayfly, they are referring to this insect,** *Ephemera danica*. **There are, however, over forty species in Britain.**

The striking metallic blue colours of the male banded agrion, *Agrion splendens*. A common damselfly of our cleaner rivers.

Common comfrey, *Symphytum officinale*. A common plant of our lowland riverbanks.

bite, living in the weeds and bottoms of chalk streams. In a river like this there would be over thirty species.

'Apart from the trout there are two other species of fish that would be feeding on all this abundance of life. One of these is the bullhead, which is a small fish that lives amongst the stones of the gravelly bed of the river. Its other name is Miller's Thumb, as its head is flattened rather like an old miller's thumb which might have been squashed between the grinding stones. These fishes occur in quite large numbers and although they are small they may be eating more of the invertebrates than the trout. The trout, however, take a lot of their food from the surface of the river so that they have a component in their diet which is not available to the bottom-dwelling bullhead.

'The other fish that is likely to be found in a river such as this is the grayling. This fish also feeds extensively on the freshwater shrimps whose numbers are so enormous there doesn't seem to be any problems of competition with the bullhead and trout.'

The thick vegetation beside the water was ideal for the insects to climb out on. It was a rich mixture of sedges, reeds, meadowsweets and willowherbs with great bunches of comfrey, *Symphytum officinale*, with their

The water from the Itchen cascading down into its natural river course from a sluice gate. The remaining water feeds a section of the ancient Itchen Navigation.

drooping cymes of flowers. The bumble bees were busy moving from one flower to the next collecting nectar and pollen.

Down amongst the more shaded leaves of some bur-reed was an amber snail, *Succinea*. Alasdair pointed out that snails were common along these chalk streams as the abundant supply of calcium helped to build their shells.

'It's unusual for them to build up to very large numbers, however, but the anglers are usually very keen on areas where there are a lot of snails, as they believe that they contribute towards the pink colour in the flesh of the trout. And certainly large chalk stream trout can come out with a very nice pink colour and a beautiful flavour.'

We were now some way along the river and in a bend, where the flow was less pronounced, we came across a new type of plant in the water. It had long strap-like leaves that eddied and flowed with the current. This was almost certainly the club-rush, *Scirpus lacustris*. Whilst we were watching these a brightly coloured damselfly came fluttering past us, patrolling to and fro along the water's edge. Its flashing metallic green colour and size told us that it was a female *Agrion*. Our attention was soon distracted by a very battered-looking large butterfly. It

settled in front of us and after a quick inspection we decided it was a Red Admiral on its last legs. These butterflies do not overwinter in Britain but are a summer migrant originating from broods that start earlier in the year in Mediterranean countries. This butterfly was the first of this species we had seen that summer. Soon many more would be arriving to mate and lay their eggs in clumps of nettles all over Britain.

Further on down, the path crossed over a weir where the river was split in two with the path following a higher stretch of river and the water which was passing through the sluice gates forming a lower river course. We asked Alasdair to tell us what was happening and why the river had been split like this.

'We've now come to a section of the river where man is interfering in a way that he has done with most rivers, but especially with chalk streams. The hatches here are allowing a limited flow of water to go off into what is the original channel of the river. And going straight on we have a flowing canal which still retains largely the same weeds as the river upstream. One of the interesting things about the Itchen is that this is perhaps the river with the oldest canal system in this country, and possibly Europe. I believe it was first modified for navigation by the

A grand old willow tree stands on the strip of land between the Itchen Navigation on the left and the natural river course on the right.

Above **A freshwater crayfish,** *Astacus pallipes*. **These large crustaceans need well aerated water and are fairly common in fast flowing chalk streams where the limey water helps to build up their thick skeletons.**

Bishop of Winchester around 1200. He had a pond built at Alresford, near the head of the river, which is still there to this day. This was created to put extra water into the river when it was necessary for navigation. At that time, Winchester was one of the most important cities in the country and the Itchen was a vital routeway up from the coast. As with many rivers this canal was brought back into use and developed considerably during the eighteenth century and was a navigable canal network with locks until the middle of the nineteenth century when it was overtaken by the railway system and fell into disuse. Luckily the Itchen Navigation, as it was called, has not dried out or been allowed to fall into ruin and still goes all the way down to Southampton. One of the benefits of these canals is that they have towpaths and where there is a towpath, it usually means that there is a public footpath, which is very much a bonus with these types of rivers as they do not generally have public access and are kept as very private places for the fishermen. Another similar navigation system is the Kennet and Avon Canal which at one time

A splendid view out across the lush water meadows that flank the Itchen. These areas are now classed as low-grade agricultural land but would have once been considered an important asset, as they could be flooded during the late winter to provide early spring growth for livestock. Today, they are an important habitat for many plants and insects as well as breeding birds, such as snipe and redshank.

linked London to Bristol. That canal, like many others, is now being gradually restored by enthusiasts as a recreational waterway.

'Looking at the growth of *Ranunculus* in the canal section you can see that there is still a reasonable flow. However, it is slowing down and if you look across to the original river course you can see that we are already noticeably higher than it. This will continue to become more marked until a little further along where we come to a lock.'

The bankside vegetation was still extremely lush and the bird life was now very noticeable. A family of swans was feeding on the weed beds just up the river and would probably soon be alongside to give us a thorough inspection. A moorhen was weaving its way between some clumps of reeds on the opposite bank, the large white patches on the underside of its tail flashing out a warning to other birds. In the *Phragmites* reedbeds between the canal and the river we could hear the vibrant song of a reed warbler which was probably setting up a territory in the dense stand. Every so often we would see a small dumpy brown bird like a great ball of

Right **Lowland rivers and canals are the haunt of this small member of the grebe family – the dabchick. Here the parent bird has just brought a small fish to the surface to feed its chick.**

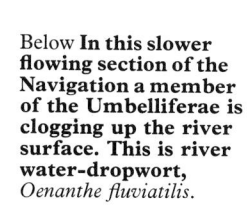

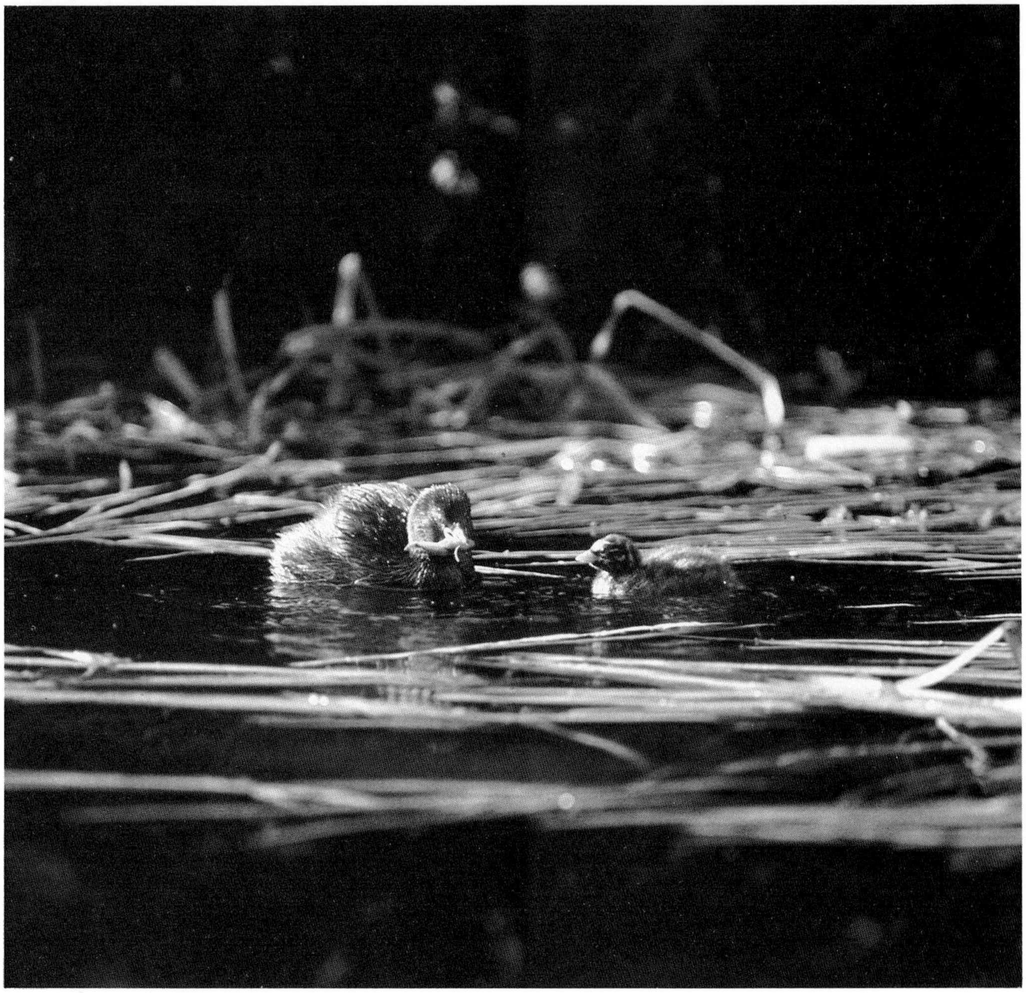

Below **In this slower flowing section of the Navigation a member of the Umbelliferae is clogging up the river surface. This is river water-dropwort,** *Oenanthe fluviatilis*.

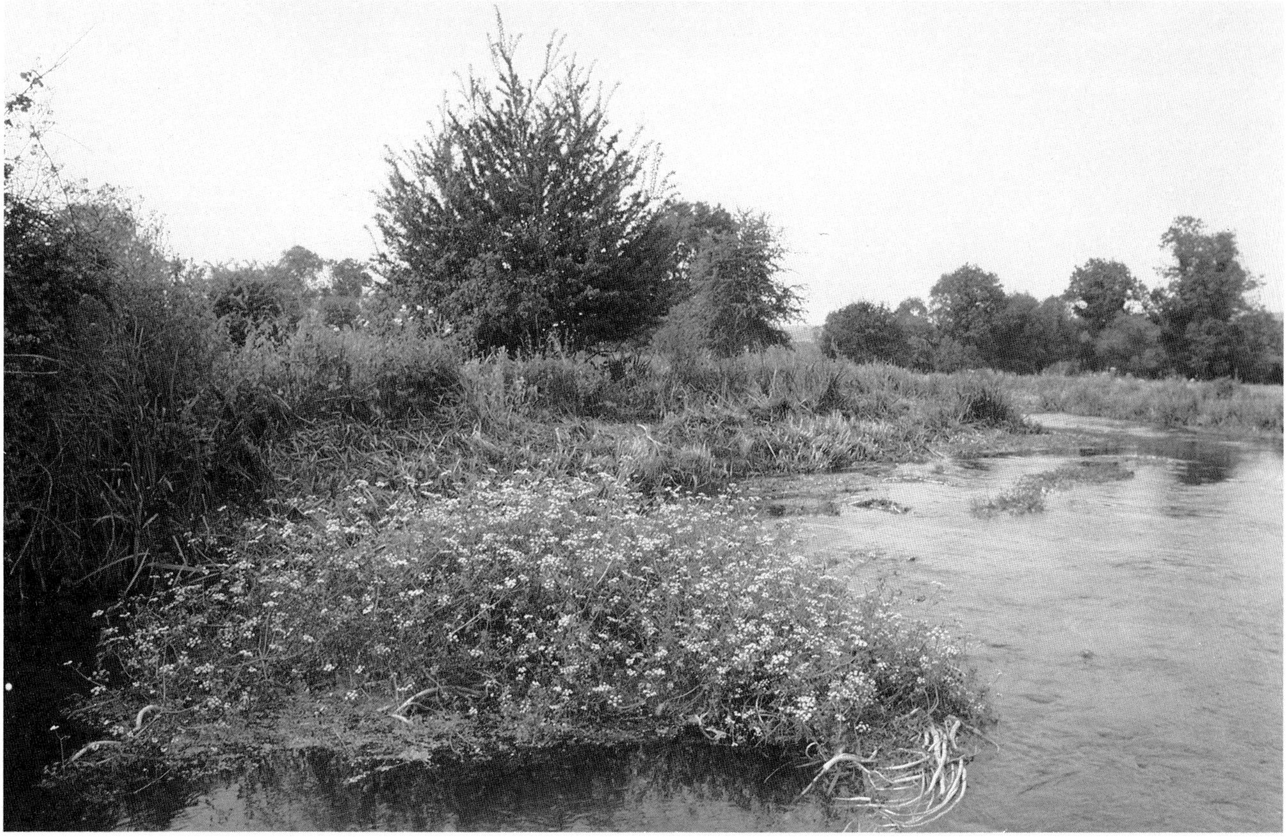

cotton wool pop up on the surface of the water and then disappear again as suddenly and as silently as it had appeared. A quick look through the binoculars confirmed that it was a dabchick, a fairly common grebe of our waterways but a rather elusive bird to watch as it dives at the slightest sign of danger. However, if you are patient and are prepared to watch quietly behind some reeds you can often obtain good views of them. In the summer you can sometimes find family groups with the striped young birds riding on the parent's back.

Out in the middle of the canal, leaves floating down stream had accumulated on a bank of flowering *Ranunculus* and amongst these an umbelliferous plant was protruding up through the mat of leaves. This was *Oenanthe fluviatilis*, a fairly common plant in the chalk streams of southern England. These aquatic and semi-aquatic umbels are often poisonous so it is best to make sure of their identification – the similar hemlock water-dropwort, *Oenanthe crocata*, which can be found along the banks here and in wet woods and marshes, is extremely poisonous.

Further along the river, again where banks of *Ranunculus* had impeded the flow, spikes of marestail, *Hippuris vulgaris*, with their distinctive whorls of leaves, were pointing up out of the water. The presence of this plant

Left **The distinctive spikes of water marestail,** *Hippuris vulgaris*, **with its whorls of leaves. This plant is found in slower flowing base-rich rivers.**

Below **A finely constructed footbridge over a channel flowing off from the main river, perhaps to feed a series of water meadows which run alongside the river.**

Two parent mute swans keep watch over their eight cygnets which are feeding on the rich harvest of river vegetation. Although the young can fly after about four months, the family party may well remain together during the winter months.

showed that the flow was considerably less than further upstream.

Around the corner at a bend in the canal there was a little low-arched bridge under the bank where a stream ran-off from the canal and across some lush meadows. Alasdair was sure that this was one of the points where water used to be taken from the main navigation to irrigate the traditional water meadow system. In order to fully explain what this was he first told us a little about the history of these ancient flood plains.

'It is thought that in prehistoric times these valleys would have been largely wooded, probably with alder carr and willow, and that the whole area was likely to have been very marshy. In some cases, there perhaps would not even have been a very distinct channel to the river. The gradient in these chalk streams is not very great and it is fairly common to find several channels in a valley today – a phenomenon known as "braiding". If you have a river which can begin to form shoals it can actually start to deviate and create separate channels for itself. At some time the tree cover would have been cut back and the river confined to definite channels so that the area for grazing livestock could be increased. The resultant meadows between the channels were highly prized land as they were extremely lush and provided excellent grazing land. The way they kept the meadows fertile was by a careful system of controlled flooding. Essentially they were flooded during the winter to keep the frost off the grass. This worked as the water from the chalk stream was always above freezing, as we noted earlier. This encouraged growth early in the year which was very important in days when winter feed was scarce and the amount of livestock you could carry over winter depended very much on how soon you could get grass for them in the spring. There were some skilled workers whose job it was to manipulate the water on the meadows, irrigating them by a series of carriers and bringing the water back out through drains.

'I believe that in those days it was used mostly by sheep which used to be taken up onto the downs during the day and brought back onto the water meadows at night. Some of the special breeds that derive from these times have the unusual habit of defecating mainly at night. This meant that the water meadows got fertilized as well! Nowadays they are regarded as low grade agricultural land and are usually only used for grazing

cattle in the summer. They are still, however, a wonderful mixture of rushes, sedges and lush herbs such as meadowsweet. But beware as they can be treacherous places to wander across, being full of remnant ditches and carriers with little bridges and hatches to hold the water back. In fact, if you look across the meadows you can see a gate standing alone, apparently with its fence missing. The gate is probably at the end of a small bridge and what you have is two meadows separated by a ditch.'

Further downstream we came across an old lock which has had hatches put in to regulate the river flow. Below the old lock the flow seemed to slow still further and the river bottom became more silty, a fact that was underlined by the large banks of starwort which were now definitely dominant over the *Ranunculus*. At this point the river was overhung by some trees and immediately the amount of plant life in the river noticeably decreased. Shading by trees can be used as a means of controlling river vegetation.

We were nearly at our journey's end but Alasdair had one more use of the river that he wanted to show us. Just a little further along, the river was again split in two with a large brick-built mill standing over one of the channels. Alasdair explained:

'Mills are another characteristic structure on chalk streams, as on many other rivers. The big difference with this type of river is that, because of the relatively constant nature of the flow, you do not have to build a mill pond. You can rely on a steady supply of water coming downstream, even in summer. Therefore although you won't find mill ponds on a chalk stream you will find mill leets or channels such as this which usually come off the river some distance upstream. Unfortunately this mill today is looking rather sad and derelict but I am glad to say that many have been renovated of late and some are still in good working order although seldom being used for their original purpose of grinding flour.'

It had been a splendid walk. Even as we stood looking at the mill a young coot tottered across a carpet of *Ranunculus* flowers beneath us. Damselflies were darting in and out of the reeds beside us, and just as we left we spotted a young jack pike lurking in the shadows of the starwort, its dappled marking blending perfectly with the patterns cast by the light. Yes, it was certainly one of those magic days and we'd been shown that chalk streams were definitely very special places.

An old lock that has been converted to house two sluice gates. Wherever one walks along our rivers one will find the signs of man's use of the river network for navigation or power.

A Broadland walk

I first came to know the Broads through Arthur Ransome's books, Coot Club *and* Big Six. *Four superb holidays then took me from its mainstreams to its backwaters. Unfortunately this was long before my interest in things natural had really developed. However, since then I have been back again and again to look at and listen to the sights and sounds of Broadland. Booming bitterns, the rustle of the reeds and the spatter of water as a coot takes to the river from its reedy nest; it is all there just as Ransome described in words. He certainly must have known the Broads well, but not as well as the owner of that familiar radio voice, Ted Ellis. Ted has lived as part of the Broadland scene for many years and has an intimate knowledge of all aspects of its wildlife. Here, he shares its secrets and with complete authority convinces even the most sceptical reader that the most amazing fact is true: the Broads are man-made.*

The bright yellow flowers of fleabane, *Pulicaria dysenterica,* **are a common sight along any damp waterway in late summer. This picture shows a wall brown butterfly,** *Lasiommata megera,* **feeding on one of the blooms.**

Information

The landscape of the Norfolk Broads, although not rich in spectacular views of hills and dales, has a more intimate beauty that, once appreciated, draws one back time and again. Its intricate network of open water, fens and dykes, linking wooded areas with vast stretches of wind blown reeds, provides a fascinating study in the processes of plant succession. For, although the Broads were at one time simply worked-out peat-diggings that became flooded as the level of the land dropped, they are now as natural in apearance, due to plant encroachment, as any lake. In the chapter on the Lake District we saw how the open water was gradually colonised first by submerged plants, then by those that had floating-leaves and long underwater stems, such as the water-lilies, and eventually by emersive perennials, which formed a steadily encroaching reedswamp. The Broads, and also the nearby Fenland districts, show a vast range of communities between this fringing reedswamp and high woodland or, more usually, drier fields.

For the naturalist, these twilight areas with their complex associations provide some of the most exciting habitats to explore. They also provide some of the most taxing management problems for conservations. For, if they are to be maintained, the natural processes that have created them have to be held in suspension. A classic example of this is some of the rich open fen communities in areas of East Anglia, such as at Wicken Fen in Cambridgeshire. The majority of the Fens has been drained and turned over to intensive arable farming. As the drainage was carried out over the centuries, the peat gradually shrank and the level of the land dropped. This has resulted in the few remaining areas of original undrained fen standing higher than the surrounding land, and in order to prevent the fen from drying out, water levels have to be artificially maintained. The land has to also be constantly managed either by cutting the reeds and saw sedge communities or by grazing. Where woodland has encroached on the drier areas this has to be cut back as well. Some of the rare fen plants such as the fen orchid and fen violet require recently exposed peat, so peat digging is also carried out in selected areas. If this work was not done some of these beautiful but rare plants would soon die out.

Sites

The Norfolk Broads is a superb area to explore and has many reserves, many of which have limited public access. However, it is always wise to check with one of the many regional guides to confirm this before planning a visit. This list also includes some other wetland areas in Britain.

Norfolk Broads:— Surlingham Broad – Norfolk Naturalists' Trust reserve with free access from the river. **Rockland Broad** – free access for boats. **Barton Broad** – Norfolk Naturalists' Trust reserve with access for boats. **Horsey Mere** – National Trust area with access for boats. **Hickling Broad** – National Nature Reserve with access for boats.

Fairburn Ings, Yorkshire. An RSPB reserve based around a flooded area resulting from the subsidence of old coal mines. It is well-known for its wealth of waterfowl and passage migrants. Access along public footpath with two public hides.

Denaby Ings, Yorkshire. A marshy area created by mining subsidence. Mostly covered by flote-grass, with a superb insect fauna and large numbers of breeding birds. Owned by the Yorkshire Naturalists' Trust and has four observation hides and a field station.

Potteric Carr, Yorkshire. An area of reed swamp, fen and open water next to the main London-Edinburgh railway. Has an impressive list of breeding and visiting birds as well as typical fen plants. The reserve was created by the Yorkshire Naturalists' Trust in collaboration with British Rail and is a splendid example of what can be done to an apparently unpromising area.

Wicken Fen, Cambridgeshire. An important traditional fen reserve owned by the National Trust with a tremendous variety of marshland insects, birds and plants. There is a 2.4 kilometre nature trail.

Holme Fen, Cambridgeshire. A drained fen reserve which is still managed in the traditional manner. Recent peat digging within the reserve and the clearing of an old decoy pond have provided habitats for plants that prefer these more open areas. National Nature Reserve.

Stodmarsh, Kent. A rich area of open water and reed beds resulting from mining subsidence. Has large numbers of breeding birds and passage migrants as well as rich reedswamp and fen communities. National Nature Reserve.

Some Broadland and fenland plants and animals to look out for:

Plants

marsh marigold, *Caltha palustris*
meadow rue, *Thalictrum flavum*
marsh pea, *Lathyrus palustris*
great willowherb, *Epliobium hirsutum*
purple loosestrife, *Lythrum salicaria*
great water dock, *Rumex hydrolapathum*
meadowsweet, *Filipendula ulmaria*
marsh valerian, *Valeriana dioica*
bog pimpernal, *Anagallis tenella*
bogbean, *Menyanthes trifoliata*
yellow loosestrife, *Lysimachia vulgaris*
wild angelica, *Angelica sylvestris*
pennywort, *Hydrocotyle vulgaris*
water-dropworts, *Oenanthe* spp.
lesser water-parsnip, *Berula erecta*
milk parsley, *Peucedanum palustre*
red rattle, *Pedicularis palustris*

fen bedstraw, *Galium uliginosum*
fleabane, *Pulicaria dysenteria*
marsh thistle, *Cirsium palustre*
meadow thistle, *Cirsium dissectum*
hemp agrimony, *Eupatorium cannabinum*
marsh helleborine, *Epipactis palustris*
southern marsh orchid,
    Dactylorhiza praetermissa
yellow iris, *Iris pseudacorus*
reedmace or bulrush, *Typha*
saw sedge, *Cladium mariscus*
sedges, *Carex*
rushes, *Juncus*
common reed, *Phragmites communis*
reed sweet-grass, *Glyceria maxima*
ragged robin, *Lychnis flos-cuculi*
cuckoo flower, *Cardamine pratensis*

Plants of open water in the Broads

Although pollution has severely affected the once widespread aquatic flora of the Broads, where it survives it is extremely rich.

yellow water-lily, *Nuphar lutea*
white water-lily, *Nymphaea alba*
hornwort, *Ceratophyllum demersum*
water-milfoils, *Myriophyllum spp.*
mare's-tail, *Hippurus vulgaris*
bladderwort, *Utricularia vulgaris*
frogbit, *Hydrocharis morsus-ranae*
water-soldier, *Stratiotes aloides*
Canadian waterweed, *Elodea canadensis*
pondweeds, *Potamogeton*
holly-leaved naiad, *Naias marina*
duckweeds, *Lemna spp.*

Some special animals to look out for in the Broads

swallow-tail butterfly, *Papilio machaon*
bittern, *Botaurus stellaris*
spoonbill, *Platalea leacorodia*
marsh harrier, *Circus aeruginosus*
black tern, *Chlidonias niger*
bearded tit, *Panurus biarmicus*
Cetti's warbler, *Cettia cetti*
yellow wagtail, *Motacilla flava*
otter, *Lutra lutra*
coypu, *Myocastor coypus*

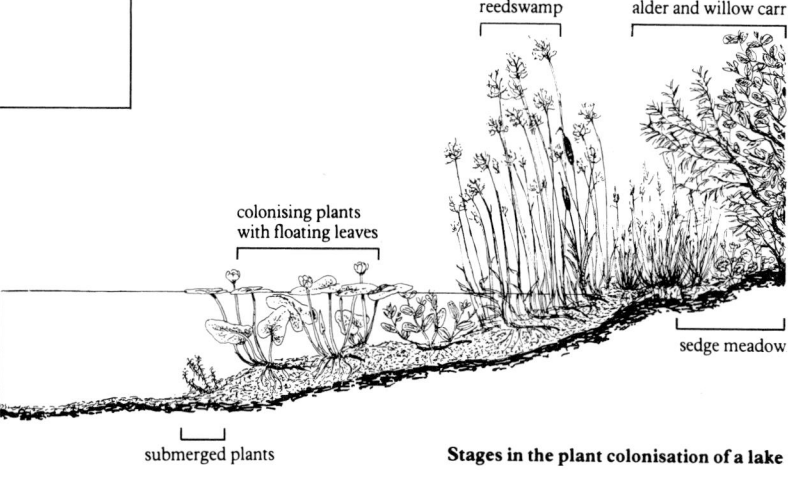

reedswamp

alder and willow carr

colonising plants
with floating leaves

sedge meadow

submerged plants

Stages in the plant colonisation of a lake

A Broadland walk
with
Ted Ellis

The Norfolk Broads are a unique and highly unusual area of wetlands transected by a patchwork of shallow lakes, lagoons, rivers and channels. They are well-known to the thousands of holiday-makers that regularly take to messing about in boats each summer and, although the area is not actually classified as a National Park, it forms as distinct and fascinating a landscape as anywhere in Britain. Naturalists have been attracted to the Broads for centuries as the rich complex of open water and fenlands provides a home for a wealth of animals and plants.

If there is one naturalist whose name is synonymous with the Norfolk Broads, it is Ted Ellis, President of the Norfolk Naturalists' Trust and a regular broadcaster both for radio and television. His book, *The Broads*, in the New Naturalists series, is considered to be the definitive account of the subject. Ted and his wife have for many years maintained their own nature reserve at Wheatfen Broad, in the Yare Valley. And it was to this haven for wildlife that we went one warm and humid day in mid-July. We met at the Ellis's converted reed-cutters' cottages and made our way out through what must be one of the most unusual gardens in Britain, for Ted's garden is quite simply a natural extension of the fen community. Here, gardening for wildlife takes on a whole new meaning. After making our way along a reed-fringed channel we stopped at a vantage point which afforded an open view of the Broad. As we took in the secluded scene, Ted told us something of the history of this extraordinary corner of the British Isles.

'The Broads are, as most people know, essentially a series of beautiful lagoons in the flood plains of several rivers running eastwards from Norwich. They spread out and meander across the low plains to eventually converge at Great Yarmouth where they empty into the North Sea. It used to be thought that they were all part of a very large estuary, which certainly existed in the area over 2000 years ago as we have found clay with estuarine shells spreading far inland underneath many of the marshes flanking the rivers. However, we know that the Broads system originated from a vast series of peat diggings that was carried on up to the sixteenth century. These have gradually been flooded over the centuries as the general water table in the region has risen. The proof confirming that this was, indeed, the case, lay in the layers of peat that have built up through the ages. If you were to sink a peat-

The dense waterside vegetation of many of the smaller Broads with their stands of reedmace, *Typha*, reed, *Phragmites*, and the great spreading leaves of the tussock sedge, *Carex paniculata*, can almost give one the impression of sailing through an Amazonian jungle.

borer deep into the ground you would find that the peat contains pollens and various fragments that enable you to date the layers back to the last ice age. When you do this across a broad you find that the natural series is suddenly interrupted in the upper portion down to about 15 feet (4.5 metres); below that the series continues at the same levels as the surrounding land. This together with documentary evidence tells us that at one time these natural-looking lakes were obvious man-made artifacts. We now believe that the practice of peat-cutting was probably imported with the Saxons and Danes when they invaded East Anglia and came to an end roughly towards the end of Elizabethan times when it became more convenient to import fuel into the region. As these pits became flooded and unworkable they provided another economy in that freshwater fish became a vital resource in the area and were sold for food in the local markets up to the nineteenth century. Wildfowl were also on the menu and even swans were cultivated as the owner of the broad had rights to these birds.

'Reed-cutting was important until the early twentieth century. But its use as an excellent thatching material has meant that it has had something of a revival in recent years. The swamp lands were often mown for a very sweet type of hay in the late summer and this practice continued until well into this century. Indeed, Broadland hay was a main source of litter for the cab horses in London during the early 1920s. In the eighteenth and nineteenth centuries, particularly following the drainage of the great fenlands, farmers turned their thoughts to reclaiming what they could of this area. They embanked the lower marshlands towards the estuary as this region had a firm raft of clay on top of the old peat deposits which was thick enough to bear the weight of the cattle and accommodate the shrinkage on drying out. These lands, which are below the sea level, are protected by high banks and drained through tidal sluices, the water being pumped out initially by wind-pumps and more recently by steam and electric pumps. With improved pumping systems the farmers are now looking to turn large areas into arable land which will inevitably produce problems for the wildlife that use these grazing levels, such as the wintering wildfowl that feed on them.

'The Broads and the system of dykes that connects them to the rivers have always been used as a means of transport in the area. The

An aerial view of Hickling Broad, one of the largest and best known of the Broads. Despite the large area of water it is rarely more than two metres (six feet) deep except in those channels dredged for sailing. The majority of the Broad is a National Nature Reserve but there is free access to river boats.

boats that were used were firstly the keels, which were small barges, and then the wherries. These were commercial single-masted boats which must have contributed considerably to the prosperity of Norwich. When the keels were removed these boats only needed a depth of four feet for navigation. In the mid-nineteenth century it is estimated that over 300 wherries were in use but with the coming of rail and road transport they began to fall into disuse and by the end of the Second World War they had all gone. Fortunately one, the *Albion*, was reconditioned by the Norfolk Wherry Trust and can still be seen plying the Broads and rivers.

'It was because of the easily accessible series of channels and open waters that the holiday industry, which is so important today, developed. People initially came here for the adventure that the maze of rivers and cuts provided as well as the opportunities for sailing in the more open water. The railways opened up the area still further and the arrival of the motor car sealed the area's fate. Today in the high season the Broads can be a very busy place but most people tend to congregate in a few key areas and therefore there are still a lot of places where you can feel that you are in a different world. After all, many people come here for the peace and seclusion that these miles of tranquil waterways can provide.

'Naturalists have always had a marvellous time here. You can go back to the seventeenth century when Sir Thomas Browne recorded the rich bird life such as the bitterns that "boomed" in the reedbeds and the different species of wildfowl, the swallow-tailed butterflies and many other creatures. Today, the rich wildlife that the early naturalists encountered is largely still here and, in fact, a great number of the broads are owned by conservation bodies such as the Norfolk Naturalists' Trust, the National Trust and now latterly the Royal Society for the Protection of Birds. What few private owners remain are all absolutely dedicated to their conservation. So that means that we can look to the future with a certain strength.

'What can a visitor hope to see? If they are patient and watch carefully, then a whole world of nature will unfold before them. There are the great tussock sedges in the swamps; the alder trees hanging over the cuts; the glimpse of a water vole heading for its hole in the bank; the families of ducks feeding in the shallows; little harvest mice

that nest in large numbers in the fens, weaving their nests about the reed stems; and, of course, in the summer there is the whispering and chattering of reed and sedge warblers, the monotonous cooing of turtle doves and the hidden chuckle of a blackcap in the sallows; a jay flapping over; the old herons dropping in at the waterside and waiting for a fish to come within shooting distance. Then there are the fish themselves. If you happen to be down in the spring, especially on a moonlit night, you can hear and see them frolicing and leaping out of the water. You might even spy a big "hen" pike making her way right into the shallows to lay her eggs. Then there are some of the birds of prey like the very rare marsh harrier that can be found nesting in the Broads and which is, happily, slowly increasing its numbers. Ospreys stop off here on their migration in spring and autumn and can be seen fishing the pools where you might catch them carrying a large bream to eat up in a tree.

'One thing that has changed in recent years that is rather sad is the pollution of the rivers and the subsequent effect on the aquatic plant life, in particular. Up until the 1950s, if you had been standing here, you would have seen green rafts of duckweeds, hornwort sprouting through to the surface and water lilies covering some of the deeper areas. With these plants there were a wealth of fresh water animals such as snails, dragonflies and beetles. The water authorities are under great pressure to control the problem and are installing a phosphate extractor at one of their sewage works on the Broads and one hopes that progress will be made. In the meantime there are still some broads that are cut off from the rivers and they will hopefully act as important reserves for the plants until the situation improves.

'Another problem that had a considerable effect on the Broads for a while was the presence of a feral population of the coypu, *Myocastor coypus*, a large South American rodent. These animals escaped from fur farms in the 1930s and found the almost limitless supplies of vegetation in the Broads very much to their liking. During the 1940s and 50s we were not aware that they would become so troublesome as they kept themselves to the dense reed swamps where they would make great nests rather like a swan's. They are mainly nocturnal feeders and were more or less left alone. But during the early 1960s their population exploded and they started to have a dramatic effect on the

Opposite **The sausage-shaped inflorescences of common reedmace or bulrush,** *Typha latifolia,* **seen here, can be distinguished from the lesser reedmace,** *T. angustifolia,* **as the male and female parts of the spike in the latter are separated by a short gap. The lesser reedmace is the more frequent of the two in the Broads.**

vegetation of the area, practically reducing marshes like these to mudflats. They also started moving out onto the arable land, eating the kale and grazing on the young corn. I can remember trapping over 2000 ourselves here in the summer of 1962 and we were not making any apparent difference to their numbers. They were running around all over the place, even around the outside of the house at night. They are quite big animals and on moonlit nights you could hear them calling to one another. I could remember that sound for quite a while. The call was rather like that of a cow. A big male coypu would be sitting on the bank opposite and you would suddenly hear the deepest threatening growl from across the water. The mothers used to sing and coo to their babies in the reedswamps. They completely changed the balance of the vegetation, in places eating back vast banks of sedges and reeds. The purple loosestrife was apparently coypu resistant and the magenta-coloured spikes of flowers appeared to bloom in profusion in the newly cleared areas. They were effectively starving themselves out. In the very cold winter of 1963 the population was severely hit and that, combined with an intensive trapping operation run by the Ministry of Agriculture and Fisheries, has managed to control their numbers. All in all the episode was a very salutary lesson in what can happen if you introduce a foreign animal into a system which has no checks on its population.

'Today as you can see the situation has returned to a fairly healthy state. If we look around we can see this open stretch of water which in this Broad is not very extensive at all. Some of them though are quite enormous, Hickling, for instance, is much bigger; you might think that you were on one of the lakes in the Lake District as it stretches away in front of you. You get the tang of the salt breeze coming in from the coast which is only a few miles away – a wonderful place for sailing. However, unlike the Lake District all the Broads are very shallow – no more than 15 feet and often as little as two or three feet in the channels. The thick layer of mud in the bottom of these areas is always building up and so the navigation authority has to dredge them from time to time. It should also be remembered that we are slowly slipping into the sea on this side of England which has been a slight help in maintaining the water levels. Also as far as the River Yare is concerned, the deepening of the bar at Great

A typical channel flanked by common reeds, *Phragmites communis*, **and reed sweet-grass,** *Glyceria maxima*. **At first glance it is hard to believe that the intricate network of channels and lakes is, essentially, the remains of a flooded medieval peat-digging industry.**

Opposite **A magnificent display of purple loosestrife,** *Lythrum salicara*.

the water are natural refuges for all sorts of hibernating insects and mammals. I've known peacock butterflies hibernate in them. They are refuges for field mice and shrews and are used as homes by stoats and weasels and it has been known for otters to nest in a whole mass of them. They were known around here as "nat-hills". I've never known exactly why, but on a warm summer's evening gnats rise up like columns of smoke from the bushes and fens. It may be to do with the use of their leaves for making little nets. We just don't know. They can go on growing until they are over four feet high but they have fairly shallow rooting systems and eventually lose touch with their water supply and keel over to start growing again. In some swamps where there are hundreds of them you can see them in all stages of decline. Sometimes whole tussock sedges were removed and trimmed to make fireside seats in the local cottages or even church hassocks. In the old days when the reed-cutters were bundling their reeds they used to cut these tough leaves and use them to tie the reeds together. Often trees will take root on them when you have a lot together and a swamp carr dominated by sallows and alder trees will take over.

'If you look behind us here you can see large quantities of these sallow bushes and alders. These have come in since the mowing of the fens has been neglected. Without the intervention of man, nature will quickly take hold of these areas and eventually the open fen communities with their rich plant-life will disappear under woodland. The traditional cutting for reeds is still continued in some broads. They are usually cut between January and April. The ones from the slightly brackish areas are the most sought after as they have the most rigid stems. They seem to thrive best in shallow water where there is some movement so that fresh supplies of nutrients can be washed in. Quite often on the nature reserves the cutting of invading bushes is done by voluntary groups and organisations such as the conservation corps which come out mostly between September and March doing a splendid job.

'One practice that is still continuing around some Broads is the cutting of the saw sedge, *Cladium mariscus*. This plant can become dominant over large areas as the evergreen leaves form dense, almost impenetrable, stands. The leaves of the plant, which are armed with sharp saw-like teeth, tend to bend over about two or three feet high

Yarmouth for shipping and the deep dredging of that river so that large boats can get up to Norwich has resulted in a tidal rise and fall affecting this whole system.

'If we look along the edge of the Broad we can see that the mud has been colonised by the tussock sedges, *Carex paniculata*, and the stands of lesser reedmace, *Typha angustifolia*, or "gladden" as it is known locally. The clumps of tussock sedge can make you feel that you are sailing down an Amazonian jungle as their spreading tufts of leaves shooting out from the top of their great pedestal, overhang the banks. The roots of these plants help to consolidate the silty bottom and eventually the accumulation of rotting vegetation forms a layer of peat, raising the level still further.

'The tussock sedges play an important part in the yearly cycle of many of the fen animals as the higher sections which remain clear of

forming a ridge which means that it is almost impossible to walk through them unless you are well protected. They are often used for the capping on thatched roofs, as they weather very well and will often outlast the reeds in the main part of the thatch. They are mown every four or five years. This is very important as they will not tolerate more frequent cutting and will soon die out to be taken over by other plants such as the black bogrush, *Schoenus nigricans*, if cut too often.

'Cutting the drier areas for hay has almost ceased in the area although some fens are grazed by horses which help to keep an excellent mixed vegetation going.'

We made a mental note to avoid heading into a saw sedge fen as we certainly were not dressed to tackle anything like that! A sudden break in the cloud allowed the sun's rays to burst through and the shimmering greens of the scene came to life. A song of a reed warbler struck up from somewhere deep in the tangle of sallows and reed serving to remind us of the wealth of wildlife hidden from view. We turned from the open water and made our way alongside the channel, crossing over small drainage ditches and holes which Ted told us were called 'pulks'. Across the channel was a magnificent display of purple flowers. These were the glorious spikes of purple loosestrife, *Lythrum salicaria*, which Ted had mentioned earlier. We stopped to admire them while Ted told us a little more.

'These tall spires of the purple loosestrife are about the height of a foxglove but look more like a buddleia except that they are a bright magenta instead of violet. July is the time to see them protruding from the river banks and around the edges of the broads. They are very attractive to butterflies such as the peacock and small tortoiseshell. Most people know that primroses have two types of flower, one kind, the 'pin-eye' with the styles at the mouth of the corolla and the 'thrum-eyed' flowers with the stamens at the mouth and the styles below. Well, in the purple loosestrife, as Darwin discovered, there are three types of flower on separate plants, all with their stamens and styles at different levels. This means that cross-pollination is, in effect, a three-way affair which is obviously a very good arrangement for the vigour of the plant race.'

Further along the channel we passed by the decaying remains of a half-submerged reed-cutters' boat – a poignant reminder of a by-gone day. However, standing on the bank

Yellow loosestrife, *Lysimachia vulgaris*, is a common plant of mid-summer along banks of rivers and lakes. Despite its name it is, in fact, a large member of the primrose family. The name *Lysimachia* (=ending strife) refers to a traditional belief that this plant placated the tempers of horses and cattle.

near it was a plant that has made something of a comeback in recent decades. Ted explained.

'This tall plant is the great marsh sow-thistle, *Sonchus palustris*, which in some ways resembles the corn or perennial sow-thistle, *S. arvensis*. The flowers are a paler yellow and it has these striking saggitate or arrow-shaped leaves that clasp the stem, which is hollow. But perhaps the most noticeable thing, if you suddenly happen upon it, is the height. They can grow to thirteen feet (four metres) in height, although between eight and ten feet is normal. They were on the verge of extinction in East Anglia at the beginning of this century. About the time of the First World War, there was just one place that was kept very secret where it managed to flourish. There then followed a great series of dredgings along the lower reaches of all the broadland rivers. This plant colonised the mud that was thrown up and is now well established almost up to Norwich on this river. It has also gone down to Suffolk and various estuarine regions. So it has gone from being almost extinct to being quite common around here.'

From here we made our way along towards a higher bank that flanked some more marshy areas. As we went along Ted enthusiastically pointed out the various plants and insects that were bordering the path.

'Here's some tufted vetch, *Vicia cracca*, with its habit of climbing up through the herbage and there is its spike of purple-blue flowers. On a nice warm evening you can pick out their very sweet almond-like scent. Paths in an area like this are most important because they provide a habitat for the shorter vegetation such as the marsh orchids and the scabious plants that can be found just here. Just along here we have the bright yellow flowers of the yellow loosestrife, *Lysmachia vulgaris*, which is actually a large member of the primrose family and not related to the purple loosestrife. Next to it is another yellow flower. This is the marsh bird's-foot trefoil, *Lotus pedunculatus*, a common plant in these marshy areas and fens.

'There are many caterpillars and little bugs and snails around, if you look. If you are simply rushing along you will not see them. But if you go regularly to a place and are patient you will begin to see so many new things. Maybe one day you will notice a spiked shield bug, *Picromerus bidens*, sucking the juice from a caterpillar, the limp body hanging down from a meadowsweet leaf. If

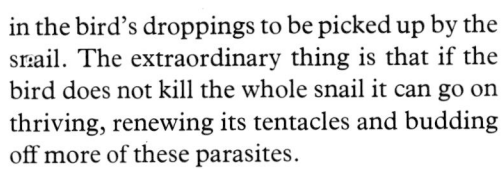

An underwater photograph of the water shrew, *Neomys fodiens.* **This aggressive little carnivore is a superb swimmer and will attempt to catch anything from insect larvae to small fish. If, while walking beside a stream or river, you hear a high pitched squealing coming from the undergrowth, the chances are that it is a member of the shrew family as apart from being noisy they guard their territories fiercely.**

you are poking around for fungi in the reed beds you might suddenly see a procession of water shrews, *Neomys fodiens,* run over your feet, squeaking.

'Here's an amber snail, *Succinea putris.* Now sometimes when you are walking through these fens and come upon one of these snails, they have what look like tiny maggots projecting from their tentacles. They can either be banded with red or green and pulsate. When I first saw them I was astonished and then I found that they were actually parts of a parasitic worm and that the pulsating was to catch the attention of water birds, like the water rail, who will hopefully eat it. The worm then becomes a parasite in the bird. At another stage it will be excreted

in the bird's droppings to be picked up by the snail. The extraordinary thing is that if the bird does not kill the whole snail it can go on thriving, renewing its tentacles and budding off more of these parasites.

'At the moment there are only a few butterflies around but in a couple of weeks these walks will be full of brimstones, whose caterpillars feed on the buckthorn bushes. The nettle beds support large numbers of peacock and small tortoiseshell butterflies. Periodically that famous East Anglian butterfly, the swallow-tail, *Papilio machaon,* will have a great flush and then they go into a very low ebb again. It is partly due to the water levels in the winter as some years the overwintering chrysalids are flooded over and drowned. If you are lucky enough to see this handsome butterfly it is quite unmistakable with its large yellow and black marked wings. The boldly marked caterpillars feed on milk parsley, *Peucedanum palustre,* which

is almost confined to this area. When alarmed they protrude an orange horn from behind the head, which gives off a curious smell, presumably to frighten predators.'

We now made our way along to an embankment that had a series of oak trees growing along it. Ted's long association with the area meant that even these trees had a definite little story.

'About 1938 the dyke alongside here was dredged and the mud from this was put on this bank. Now, what almost certainly happened was that jays brought acorns from the nearby wood and hid them in the bank and so here we are today walking underneath the shade of these 45-year old oaks as a result of the acorns that the jays failed to retrieve!

'So often in order to understand why certain plants are where they are, you have to go back to changes in land use over the years. For example, a common plant that exploits a sudden upheaval or even a fire is the rose-bay willowherb or fireweed, *Chamaenerion angustifolium*. This plant with its wind-blown seeds will quickly exploit the resultant bare ground. An interesting case here was a patch in the fen that used to support a lot of marsh orchids. I had always wondered what was so special about that particular area and then one day one of the old marshmen told me that he used to pen his geese on that land. The effect of their grazing and droppings had therefore lasted for years after they had gone. Again we keep returning to the effect of man on the balance of the community. Another instance, which indirectly illustrates this, is the bank. In the 1940s it used to have enormous numbers of orchids growing on it and at that time the place was overrun with rabbits which helped to keep the vegetation short. In 1954 myxomatosis struck and all the rabbits were killed. We didn't see any for five years and rapidly all the lower growing plants began to be smothered. When one keeps detailed records over the years you can see just how dramatic these changes can be.

'Looking out from here we can see a stretch of fen with a lot of meadowsweet and meadow-rue, behind that are hundreds of bushes of the guelder rose, *Viburnum opulus*. These are crimson in the autumn with the fruits which seem to attract birds even through the fog; the red lamps of the clusters of fruits and leaves obviously penetrate the gloom. I have seen fieldfares and wood-pigeons flying out of the thick mist to alight on them. In some years we have flocks of waxwings from the continent feeding all over them.

'Beyond this stretch the path goes into a rush fen with masses of small rushes and milk parsley together with marsh cinquefoil and cottongrass. Further out we reach a lower fen with a great deal of saw sedge and bog myrtle or sweet gale, *Myrica gale*. This shrub is usually dioecious, that is it has male and

Two mating amber snails, *Succinea putris*. The Broads have a rich invertebrate fauna which is not always appreciated by the many holiday-makers, but patient study will reveal a fascinating world.

female flowers on separate plants, but many years ago I came across a solitary large bush which had both male and female flowers on it. At the time I examined hundreds of others in the same marsh but could not find another instance of this. It was recorded back in the 1930s that very old bog myrtles will sometimes partly change their sex. And now the bushes are generally a good deal older I have been able to find more instances of this.'

We now headed back towards some woodland that was fringing the marsh. Along the path the way was blocked by the fallen trunk of a large tree. Ted pointed out that it came down with many other trees in January 1976 during a gale.

'We've kept a few like this one so that fungi will develop on them. This is a beech and there is an oak further along. When oaks fell in the old days they were gradually covered with new peat and became preserved as "bog-oaks". This one has only been down for seven years and already it is disappearing. First of all it has this great covering of moss which absorbs moisture from all the marsh mists. The woodlice live in the rotting bark and break it down forming a fine soil for the moss and other plants to root in. Already we have a whole list of plants growing on it – ragged robin, a little burdock, there's some bindweed, woundwort, a vetch, nettles, the inevitable willowherb and some hemp agrimony. The whole thing will become spongy and will soon break down. It is rather fun to see exactly what is happening.

'Here at the edge of the wood we have many alders growing. Often you will come across alders and willows with several trunks sprouting from one base – the result of coppicing by man. When you see them singly

it means that they have never been cut. Here is an example of this – a very big alder tree and further along we have some tall ash trees – gradually we will see that it builds up to an oak-ash woodland with some hazel in the understorey. This type of woodland has a long history in the area as deep peat bores have been made down to over 40 feet (1250 centimetres) and it has been found that there are signs of hazel, for example, at intervals right down to layers which are over 8000 years old.

'One of the things I find of particular interest is the decaying litter, here in the swamy fens. Because waters keep flooding in due to the tidal effect it is always spongy and wet and there are few air spaces in the accumulated material so that it does not decay all that quickly unlike some of the more efficiently drained fens, such as Woodwalton and Wicken fen in Cambridge. These drier fens are not subjected to the continual wetting and drying and consequently millipedes flourish and reduce the litter to crumbly, black, granular peat within a comparatively short time. Here, however, the slowly decaying mats afford a refuge for an enormous number of most interesting little fungi – micro-fungi and insect fungi. In fact, this is the richest area in the whole of Britain for this fantastic miniature world.'

We made our way through the wood and out into a field and started up the edge to some higher ground where we would have a better view of the Broad. As we walked up we asked Ted how a visitor to the area might best explore it.

'Access to the Broads is mainly by water but they do have miles of these flood banks with walks along them in the lower reaches – at Hickling, for instance. Otherwise access consists largely of little footpaths which originally ran from village to village along the fringes of the river valleys. Some of these which end at the river are called "lokes". They will take you down so you can glimpse the river and maybe run to a little riverside pub. The main advice is to buy a good map which shows all the footpaths. Also it is a good idea to try and find out a little about the history of the area you are visiting. Some of the little tracks and the marshes around them tell a story of quite ancient history. For example, a short distance from here there is a track known as "Smee Loke" which leads to what is now deep marsh. That is the "Smee" or "Smeeth" which, in Saxon times, would have been a hay or possibly a grazing

An orb-web spider, *Araneus,* at the centre of its web. This often malign group of invertebrates find a rich harvest amongst the reed beds and river sides.

meadow. One does find the bones of Celtic short-horned cattle and sheep in marshes where they could not possibly feed today. Sometimes they are as much as eight feet (2·5 metres) down in the peat. This is obviously evidence that the land was higher in Saxon times than in Roman times when, as we have noted, a lot of the area was under an estuary. Now we are going down again. Even old maps and documents give a glimpse of the effects of these changes. The field names can add to the story. For example these fields around here all had tree names – the great birch, the little birch, etc. – indicating that they were formerly part of an extensive woodland that has long since been cleared.

'The enquiring mind can always add so much more to a ramble by finding out about special areas in the local museums and libraries, by getting to know people who have similar interests and meeting local people. Always after your ramble you should have all sorts of questions in your mind that need answering. There are few things as satisfying as piecing the story of our landscape together and then applying it to help further understand the natural history.'

By now we had reached a point where we had a clear view of the nearby Broads. We followed Ted's gaze as he looked back out over the area we had just explored and beyond. In a larger Broad a few pleasure cruisers and sailing boats had discovered this lovely area. If the passengers saw only a tenth of that which we had seen that day they would have been as delighted as we were with this most enthralling landscape.

Our day had come to an end. Ted Ellis had given us a privileged glimpse into the Broads. We can only hope that the quiet efforts of individuals like Ted and bodies such as the Norfolk Naturalists' Trust will continue to ensure the future of this marvellous corner of England.

Further reading

Bang, P. and Dahlstrom, P., *Animal Tracks and Signs*, Collins (1974).

Bellamy, D. J., *Bellamy on Botany*, BBC Publications (1972, revised 1975).

Bellamy, D. J., *Bellamy's Britain*, BBC Publications (1974).

Bellamy, D. J., *Botanic Man*, Hamlyn (1978).

Brown, R. W. and Lawrence, M. J., *Mammals of Britain*, Blandford (1967, revised 1974).

Bruun, B. and Singer, A., *The Hamlyn Guide to Birds of Britain and Europe*, Hamlyn (1970, revised 1978).

Chinery, M., *A Field Guide to the Insects of Britain and Northern Europe*, Collins (1973).

Clarke, P. *et al*, *The Sunday Times 1000 Days out in Great Britain and Ireland*, Macdonald (1981).

Clegg, J., *The Freshwater Life of the British Isles*, Warne (1965).

Corbet, G. B. and Southern, H. N. (eds.), *The Handbook of British Mammals*, Blackwell (1977).

Darlington, A., *The Pocket Encyclopaedia of Plant Galls*, Blandford (1968, revised 1975).

Ellis, E. A., *The Broads*, Collins (1965).

Haslam, S. M., *River Plants*, Cambridge University Press (1978).

Haslam, S. M., Sinker, C. A. and Wolseley, P. A., *British Water Plants*, Field Studies 4 (1975).

Holliday, F. G. T., *Wildlife of Scotland*, Macmillan (1979).

Hubbard, C. E., *Grasses*, Penguin (1968).

Humphries, C. J., Press, J. R., Sutton, D. A., *The Hamlyn Guide to Trees of Britain and Europe*, Hamlyn (1981).

Mabey, R., *The Common Ground*, Arrow (1981).

Macan, T. T. and Worthington, E. B., *Life in Lakes and Rivers*, Collins (1951)

Measures, D., *Bright Wings of Summer*, Cassell (1977).

Morris, P. (ed.), *The Natural History of the British Isles*, Country Life (1979).

Ogilvie, M. A., *The Birdwatcher's Guide to the Wetlands of Britain*, Batsford (1979).

Phillips, R., *Grasses, Ferns, Mosses and Lichens of Great Britain and Ireland*, Pan (1980).

Pollard, E., Hooper, M. D. and Moore, N. W., *Hedges*, Collins (1974).

Rose, F., *Wildflower Key*, Warne (1981).

Whalley, P., *Butterfly Watching*, Severn House (1980).

Organisations to join

Botanical Society of the British Isles
68 Outwoods Road, Loughborough, Leicestershire.
A national society for both amateur and professional botanists. Organises mapping schemes and is active in the conservation of our wild plants.

British Butterfly Conservation Society
Tudor House, Quorn, Leicester.

British Trust for Conservation Volunteers
10–14 Duke Street, Reading, Berkshire RG1 4RU.
An organisation for people over sixteen years of age which undertakes practical projects, such as clearing scrub, maintaining reserves, tree-planting, etc.

British Trust for Ornithology
Beech Grove, Tring, Hertfordshire.
National organisation which carries out research into all aspects of bird life supported by a growing army of amateur enthusiasts.

Ramblers' Association
1–5 Wandsworth Road, London SW8 2LJ.

Royal Society for Nature Conservation
The Green, Nettleham, Lincoln.
The Royal Society for Nature Conservation is the national association of the 44 local Nature Conservation Trusts which form the major voluntary organisation concerned with all aspects of wildlife conservation in the United Kingdom. The Trusts have a combined membership of 140,000 and, together with the Society, own or manage over 1,300 nature reserves throughout the UK covering a range of sites, from woodland and heathland to wetland and estuarine habitats. Most Trusts have full-time staff but the members themselves, with a wide range of skills, contribute greatly to all aspects of the work.

Royal Society for the Protection of Birds
The Lodge, Sandy, Bedfordshire.
The major conservation organisation for birds and their habitats.

The Scottish Wildlife Trust
25 Johnston Terrace, Edinburgh EH1 2NH.
The Scottish branch of the County Conservation Trusts.

Watch: The Watch Trust for Environmental Education
22 The Green, Nettleham, Lincoln LN2 2NR.
Sponsored by *The Sunday Times* and the *Royal Society for Nature Conservation;* WATCH is a national club for children and young teenagers.

Index

Figures in italics refer to illustrations